ANNALS OF MATHEMATICS STUDIES
NUMBER 5

The Two-Valued Iterative Systems
Of Mathematical Logic

BY

EMIL L. POST

PRINCETON
PRINCETON UNIVERSITY PRESS
LONDON: HUMPHREY MILFORD
OXFORD UNIVERSITY PRESS

1941

PRINTED IN U.S.A.
Lithoprinted by Edwards Brothers, Inc., Lithoprinters
Ann Arbor, Michigan, 1941

Dedicated

to

CASSIUS J. KEYSER

in one of whose pedagogical devi-
ces the author belatedly recognizes
the true source of his truth-table method.

CONTENTS

CONTENTS

INTRODUCTION

In its original form the present paper was presented to the American Mathematical Society, April 24, 1920, as a companion piece to the writer's dissertation, henceforth referred to as Elementary Propositions [23][1]. Though submitted for publication the following year, at least a revision of the present paper was editorially suggested. Except for the finishing touches, including therein introduction and footnotes[2], the present version was effected in the years 1929-1932.

Though, as suggested by its title[2a] the main contribution of the present paper may be considered to be the complete solution of a certain problem in classical mathematics, the origin of the paper and, as far as the writer can see, its main interest lies in the study of two-valued propositional calculi. In the first of the four parts into which Elementary Propositions was divided the propositional calculus of Principia Mathematica [37], based on the primitives negation and disjunction, was studied; and, as a preliminary to that study, it was shown that if what we termed truth-tables were assigned to ~p and pvq in accordance with their heuristic meanings, then not only was a truth-table thereby formally determined for every expression in the calculus, but every (two-valued) truth-table was thus obtained. The second part of that paper attempted to generalize the first part by allowing for an arbitrary finite number of primitive truth-functions, of arbitrary finite numbers of arguments, to each of which a truth-table was arbitrarily assigned. It was then immediately obvious that while each expression in such a calculus again had a corresponding truth-table, the set of such truth-tables might

---

[1]   Numerals in brackets refer to the bibliography concluding the present paper.

[2]   However, footnotes 47 and 48 (§25) were also written within the period stated.

[2a]   This rather refers to theroriginal title of the present version, "Determination of all iteratively closed two-valued systems of functions."

now be but a proper subset of the set of all possible truth-tables. Two problems thus suggested themselves. First, what truth-tables could be assigned to the primitive functions so that the set of all possible truth-tables would result; secondly, allowing for arbitrary initial truth-tables, what are the various possible sets of truth-tables that can thus result. Added interest was lent to the first problem by Nicod's achievement [21] in basing a propositional calculus on but one primitive truth-function in accordance with the pioneering work of Sheffer [27][3]. While the second problem was at first viewed as an aid to the solution of the first, its intrinsic interest soon became apparent, and made it our main problem.

A brief summary of the result of our solution of this main problem appears in <u>Elementary Propositions</u>[4]. In this connection it should be noted that the operation of substitution, which is the only operation used in building up the various expressions of a propositional calculus by means of variables and primitive functions, has its counterpart for the corresponding truth-tables. The problem can then be considered entirely in terms of truth-tables. The above summary is then at the moment best restated as it appears in the printed abstract of the present paper [22] "......there are 66 different systems generated by primitive tables with no more than three arguments, and 8 infinite families of systems which require tables of four or more arguments." Such a generated system of truth-tables clearly has the property of being closed under the operation of substitution, or, as we shall say, iteratively closed. It is then significant, as was noted in our abstract, that the converse also is true.[5] In

---

[3]   On the other hand, such a basis for a logic of propositions as given by Tarski [30] is outside the scope of our investigation; since, in addition to the primitive truth-function equivalence, it employs quantification of propositional variables.

[4]   See [23], §5, (pp. 173-4).

[5]   Provided we exclude the null system of functions, which vacuously satisfies the condition of iterative closedness, or else allow a null set of generators, thus adding one to the number of generated systems. While this omission is easily supplied (but see footnote 51 (§25)), more serious is our exclusion of functions of no variables, i.e., constants. In this omission we follow the usual form of a propositional calculus, e.g., such as that of <u>Principia Mathematica</u>. Actually, there would be little difficulty in modifying our count of systems in

terms of generated systems and logic, we may be said to have determined all the non-equivalent sub-languages of the language of
the complete two-valued propositional calculus. Since in itself
a truth-table is but a two-valued function, two-valued referring
to the range of values of arguments and possible values of function, in terms of iteratively closed systems we are led to the
title of this paper,[5a] or, to use the terminology of Wiener,[6] we
have determined all iterative fields of two-valued functions.

The first version of the present paper was presented in
terms of truth-tables, and followed the method of discovery via
the concept of generated system.[7] That every closed system of
truth-tables could be generated by a finite set of tables only
appeared after all such generated systems had been found. Since
it was impossible to present all the calculations and tabulations, in addition to arguments, inherent in this procedure,
that first version may be said to have been but a report on the
complete solution. The present version centers around the concept of closed system, and merely pauses in each instance to
verify that the closed system in question can be generated by a

---

the two-valued problem to allow for the two constants there occurring, since those constants play essentially the same rôle as
two of the four two-valued functions of one variable. On the
other hand, generalizing our membership-functions of classes,
(§2), to allow for arbitrary constants would fundamentally alter
the scope of the present paper, as well as break down the iso-
morphism with the two-valued problem.

Our count of systems would have been considerably simplified if we failed to distinguish, for example, between functions
of one variable and functions of two variables whose values are
independent of the value of one of those variables. That such
distinctions are real can be illustrated in a variety of ways;
and while the simplification of the count, if desired, can be
immediately effected, not so the meticulous recovery of the lost
distinctions.

5a   See footnote 2a.

6   See [41], p. 7, for the definition of iterative field, [40],
p. 159, for the class of 'operations' generated by a single operation. Where Wiener analyses the operation of substitution,
which is at the bottom of these concepts, into at least semi-
atomic parts, we have found it more convenient to use a more
uniform definition. On the other hand, Wiener makes no reference to a specific class of variables to be employed, thus making the first of the above definitions ambiguous, the second
misleading.

7.   A brief outline of this method might be given. Since, neg-
lecting the particular variables heading a truth-table, there
are but four truth-tables of one argument, sixteen of two, while

finite set of generators. Its synthetic development unifies the
argumentative portions of the original development, and avoids
all of the latter's tabulations, and most of its calculations,
so that it is complete in itself.[8] Furthermore, following the
suggestion of the referee of the original version, the truth-
table has been replaced by logical expressions in the Jevons notation. To preserve the precision of the truth-table development
these expressions are interpreted not as truth-functions of propositions but as what we call membership functions of classes.
While the advantages are not all with the Jevons notation, its
greater flexibility perhaps justifies the change.[9]

   To avoid the unclarity of the first version of the present
paper, we have devoted the first two sections of the present

---

the tables of one or two arguments generated by a set of tables
similarly restricted can be found by using only such tables,
mere systematic calculation and tabulation served to yield all
the distinct sets of such tables that could be so generated. On
the other hand, special methods adapted to the individual cases
were needed to obtain a formula for an arbitrary table in each
of the corresponding systems. The next step was a systematic
procedure which, again by special arguments, yielded formulae
for all tables of more than two arguments which, taken singly,
would generate systems not identical with one of the previously
found systems. By restricting these tables to those of three arguments, all systems that could be generated by one table of
three arguments, but not by tables of one and two arguments, were
found, whence it was an easy matter to find all systems generated by arbitrary sets of tables of no more than three arguments.
One repetition of this weeding-out process then so cleared the
field that it was possible to follow the procedure for 'third
order' systems in the case of n-th order systems with arbitrary
n > 3, and thus complete the solution.

[8]   On the other hand, much additional information yielded by
the original solution is lost. But see footnote 50, (§25).

[9]   Particularly for those systems satisfying the '[A:a] condition', and thus admitting for their members the 'normal [A:a]
expansion', (see §8). It should be mentioned that in 1920-21
Professor Oswald Veblen urged the writer to replace the truth-
table solution by one which represented logical functions by
algebraic polynomials modulo 2 (the arithmetical forms of B. A.
Bernstein [2], whose importance has since been emphasized by
Stone [28] in connection with his researches on Boolean algebras. For their connection with Veblen see [32], P. 9). However, it then seemed to the writer, and still seems so, that
while the derivation of what we have called the alternating systems (§15) would be greatly simplified thereby, all of the rest
of the solution would become correspondingly more complicated.

version essentially to the statement of our problem. The next
three sections develop those aspects of the technique of the
Jevons notation which are required in the solution of the problem. The last four sections of Part I form an integral part
of that solution, but are included among the preliminaries to
expedite the presentation of that solution.

Part II is devoted to the complete solution of what we have
called our main problem. The various closed systems of membership functions of classes are presented either individually, or,
in the case of the infinite families of closed systems, with aid
of a parameter. In general we may say that each closed system
is given by a condition on its members. These conditions are
such that mere inspection is sufficient to tell of a given expression whether the corresponding function satisfies that condition. They are easily retranslated in terms of truth-tables.

The first of the two sections comprising Part III unifies
the set of all closed systems of membership functions, or truth-
tables, with respect to the relation of inclusion. By analysing
this relation in terms of immediate inclusion, we are led to the
inclusion diagram, which gives a graphical summary of the solution of our main problem. It should be noted that even this re-
vised version of the section was written before the christening,
and recent development, of the lattice concept.[10] The last section gives our solution of the other of the two problems posed
above. This problem was also solved in the first version of the
paper under the added condition that the generators be independent, and then further specialized in such a way that the infinite number of independent sets of generators reduced to 36 essentially different sets. Actually, in the light of the solution of our main problem, the solution of the secondary problem
without the condition of independence is shown in the present
version to be very simple. A different and far easier solution
than in the first version is here obtained of the problem of independent generators, but with corresponding added complications
in the derivation of the specialized result. That the same 36
essentially different sets are then obtained is but a check on
both solutions.

---

[10]   See for example, [5].

The writer is aware of but two papers directly touching on our two problems. Both papers restrict themselves to second order functions, i.e., functions of but two variables, not only as generators but as functions generated. There being but sixteen such functions, the doubly infinite character of our problem thus becomes completely finite, and manageably so. In 1925, E. Zylinski [42] specifically determined the sets of second order functions generated by single second order functions. In a recent investigation, W. Wernick [36] discusses the sets of second order functions generated by arbitrary sets of second order functions, and specifically determines what we would call the various sets of second order functions which are sets of independent generators of the sixteen second order functions, and hence, indeed, of all functions. Since our general solution of the problem of independent generators can but be given by conditions on those generators, while our specialization rules out certain sets even consisting of second order functions, these results may be said but to overlap our own. We might note that had Zylinski seen our Elementary Propositions, published four years earlier, he would have found the question ending his paper answered in the affirmative, the conjecture immediately preceding disproved.

In the second part of Elementary Propositions will be found the following paragraph. "We thus see that complete systems are equivalent to the system of 'Principia' not only in the truth table development but also postulationally. As other systems are in a sense degenerate forms of complete systems, we can conclude that no new logical systems are introduced". Perhaps the essentially new logics yielded by our postulational generalization blotted out our earlier idea that these subsystems of the complete system would constitute a sort of logical playground for the construction of what are now called propositional calculi. This earlier vision was renewed in the observation that such a system as that generated by implication, if admitting a postulational development making it (postulationally) 'closed', would not be 'completely closed' — to use concepts of the third part of Elementary Propositions. Actually, quite a number of these systems should admit a postulational development in the ordinary sense. And, indeed, through developments of Łukasie-

wicz, Tarski, Bernays, Lésniewski, Mihailescu, and Wajsberg, four
of these systems may be said to have been so treated. In particular, we may refer to developments in terms of equivalence
only ([14], [33], and [16]),[11] which generates the system we
symbolize L_2, (§15), equivalence and negation [17], and equivalence and its dual [18], each pair of which functions generates
L_1, (§15),[12] implication ([34], [10]), which generates F_4^∞,
(§22), and implication and conjunction [10] and perhaps equivalence and disjunction [19], among others, which generate the important C_2, (§19).[13] We might add that where an ordinary postulational development is not possible, a complete logic might
be given for a system in other ways.[14]

While our main problem and its solution, retranslated in
terms of truth-tables, fits hand in glove with developments of
two-valued propositional calculi based on a finite number of
primitive truth-functions, difficulties in formulating a concept

---

[11]   In connection with these developments, and also the two
immediately succeeding, see [24].

[12]   The reviewers' observation that not all enunciations in
terms of these primitives capable of proof in ordinary logic
can be deduced here makes it questionable whether we can say
that the postulates are for system L_1. Though here it may be
just a case of incompleteness, in other instances it may be
that not the two-valued logic is under consideration.

[13]   We may say that as system C_1 is the concern of the complete two-valued calculus of propositions, system C_2 is the
concern of the 'positive logic', ([9], p. 68). In this connection C_3, the dual of C_2, is distinguished by the fact that
under the functions of classes interpretation it consists of
all such functions whose values, for given classes, are independent of the particular universal class containing the given
classes. Returning to the postulational developments, we must
admit that the intent of the authors is not to develop a postulational development for a given system via certain primitives, but merely to study the inter-relations of those primitives among themselves. Thus, in [10] again, we find a study of
expressions built up by implication, conjunction, and negation,
while either of the first two primitives with the last generate
the complete two-valued system.

[14]   Thus, in the duals of the first class of systems a 'contra-
dictory logic' is possible (see [11]), while in all cases it may
be possible to set up a complete logic solely by means of rules
of derivation (see [20]).

of identity of truth-functions, as against mere equivalence, prevents us from giving even a definitive formulation of the problem of determining all iteratively closed systems of truth-functions. That our solution of the corresponding problem for truth-
tables may here be of little help is indicated by the following
specialization of the general problem. In the propositional
calculus of <u>Principia Mathematica</u> let two truth-functions be considered identical when and only when their expressions in terms
of the primitives $\sim p$ and $p \vee q$ are identical. The concept
of an iteratively closed system of $(\sim, \vee)$ truth-functions is
then precise. But while to each closed system of $(\sim, \vee)$ truth-
functions there corresponds the closed system of corresponding
truth-tables, the problem of determining all such closed systems
of truth-functions is obviously entirely independent of the corresponding problem for truth-tables, being a problem in symbol
structure only. A solution of this problem in the same sense as
that of the present paper is hardly to be expected, since the
cardinal number of $(\sim, \vee)$ closed systems is that of the contin-
uum.[15] Indeed, we hazard the guess that in some sense the problem is unsolvable. In particular, it seems likely that the problem of determining for any two finite sets of $(\sim, \vee)$ truth-functions whether the systems they generate have a function in common is unsolvable in the sense of Church [7]. We may finally
note that unlike iteratively closed systems of truth-tables, an
iteratively closed system of $(\sim, \vee)$ truth-functions need not be
capable of generation by a finite set of generators.[16]

---

[15]   In fact, even those consisting of functions of one variable.
For obviously such a function may have an arbitrary number of occurrences of that one variable. Furthermore, if $f(p)$ has m
occurrences of $p, g(p), n, f[g(p)]$ has mn occurrences of p. It
follows that for any given subset, finite or infinite, of the set
of primes $2, 3, 5, 7, \ldots$ the set of $(\sim, \vee)$ functions of one variable for which the number of occurrences thereof factors into
primes in the subset is iteratively closed.

[16]   This follows from the number of $(\sim, \vee)$ truth-functions being denumerably infinite, of closed $(\sim, \vee)$ systems non-denumer-
ably infinite. In particular, each of the closed systems of the
preceding footnote corresponding to an infinite number of primes
is incapable of being generated by a finite set of generators.
For a finite number of such functions will, by the reasoning of
that footnote, at most generate a system corresponding to a finite number of primes. In the terminology of §1, we may thus say
that not every iteratively closed $(\sim, \vee)$ system is of finite

On the other hand, the m-valued logics of the fourth part of Elementary Propositions do suggest a natural extension of our present problem. Added interest is lent to such a possible extension by the fact that the m-valued logic of Łukasiewicz and Tarski seems to correspond to but a proper subset of the set of all possible m-valued truth-tables.[17] In particular, for three-valued logics the corresponding problem of determining all iteratively closed three-valued systems of functions should be a feasible one. The immediate chances for a complete solution of the m-valued problem are small, for, as has been observed by Wiener in different phraseology [40], among the iteratively closed systems of m-valued functions of one variable are included all substitution groups of degree m.[18] However, the various conditions defining the iteratively closed two-valued systems of our solution easily generalize, and in more than one way, to give certain iteratively closed m-valued systems. And it may be feasible to obtain a general solution of the m-valued problem for arbitrary functions in terms of the corresponding problem for functions of one variable.[19]

---

order. This continues to be so even with 'finite order' generalized in the sense of footnote 24 (§1). For the set of all $(\sim, v)$ functions of the form $\sim P$ is clearly iteratively closed. By letting P involve v's only, we can have such a function on an arbitrary number of arguments with but one occurrence of $\sim$. On the other hand, such functions of n or fewer arguments can only generate functions of more than n arguments which involve at least two $\sim$'s.

[17]   See [35].

[18]   The substitution groups correspond rather to the contracted closed systems (see §1). In the same sense, the iteratively closed m-valued systems of functions of one variable in their entirety are identical with the generalized substitution groups of degree m in the sense of Suschkewitsch [29].

[19]   In this connection, see §10.

Part I

PRELIMINARIES

§1

ITERATIVE CLOSEDNESS AND GENERATION

We briefly term <u>function</u> what Whitehead and Russell call 'the ambiguous value of a function' which, in turn, is essentially the 'dependent variable' of mathematics. We postulate a class of independent variables in terms of which each function is understood to be ultimately expressed. A particular method of defining a function of a given set of independent variables will be called an <u>operation</u> on those variables. Functions yielded by different operations on the same set of independent variables are then considered to be identical if for all admissible values of those variables the corresponding values of the functions are always identical.

Let V be a denumerably infinite class of independent variables $x_1, x_2, x_3, \ldots$, and let F be an existent class, or, as we shall say, system of functions whose arguments, finite in number and existent for each function, are members of V.[20] Let the variables in V range in value over one and the same class of values v, and let the corresponding values of the functions in F also belong to v. If $X_1, X_2, \ldots, X_n$ represent variables in V or functions in F, those variables and functions will in their totality depend on a finite number of distinct variables $x_{j_1}, x_{j_2}, \ldots, x_{j_m}$, belonging to V. If then $f(x_{1_1}, x_{1_2}, \ldots, x_{1_n})$ is also a function in F, on distinct variables $x_{1_1}, x_{1_2}, \ldots x_{1_n}$, $f(X_1, X_2, \ldots, X_n)$ will represent a function of $x_{j_1}, x_{j_2}, \ldots, x_{j_m}$. This function will be said to be obtained from $f(x_{1_1}, x_{1_2}, \ldots x_{1_n})$ by replacing $x_{1_1}, x_{1_2}, \ldots, x_{1_n}$ by $X_1, X_2, \ldots, X_n$ respectively. And the expression $f(X_1, X_2, \ldots, X_n)$, which not only represents this function but indicates the manner in which it is to be obtained, will be referred to as an <u>iterative process</u> depending on the variables $x_{j_1}, x_{j_2}, \ldots, x_{j_m}$ and yielding the function in

---

[20]   See footnote 5 (Introduction).

10

question. We now define F to be <u>iteratively closed</u> with respect to V if replacing the arguments of a function in F by variables in V or functions in F always results in a function in F, i.e., if whenever $f(x_{1_1}, x_{1_2}, \ldots x_{1_n})$ is a function in F on distinct variables $x_{1_1}, x_{1_2}, \ldots x_{1_n}$, and X_1, $X_2, \ldots X_n$ represent variables in V or functions in F, then the function yielded by the iterative process $f(X_1, X_2, \ldots X_n)$ is also in F. A system of functions iteratively closed with respect to V will also be referred to as an iteratively closed system over V.[21]

A function of n distinct independent variables will be said to be of order n. We then immediately verify that an iteratively closed system F either consists wholly of first order functions or else possesses functions of every finite order.

For if F does not consist wholly of first order functions it will possess some function $f(x_{1_1}, x_{1_2}, \ldots x_{1_m})$ of order m > 1. Then, by the closedness definition, the first order function $f(x_{j_1}, x_{j_1}, \ldots x_{j_1})$ belongs to F, and if the n-th order function $g(x_{j_1}, x_{j_2}, \ldots x_{j_n})$ is in F, and $x_{j_{n+1}}$ is distinct from $x_{j_1}, x_{j_2}, \ldots x_{j_n}$, $f[g(x_{j_1}, x_{j_2}, \ldots x_{j_n}), x_{j_{n+1}}, \ldots x_{j_{n+1}}]$ will be an $(n+1)$-th order function in F. The result follows by mathematical induction.

Our proof actually shows that if a closed system F does not consist wholly of first order functions, then it possesses at least one function of each finite set of variables belonging to V. Now if $x_{1_1}, x_{1_2}, \ldots x_{1_n}$, and $x_{j_1}, x_{j_2}, \ldots x_{j_n}$ are any two sets of n distinct variables in V, then, if $f(x_{1_1}, x_{1_2}, \ldots \ldots x_{1_n}$ is in F, $f(x_{j_1}, x_{j_2}, \ldots x_{j_n})$ will also be in F, and conversely. It easily follows that the totality of functions in F on the first set of n variables is transformed in 1-1 fashion into the totality of functions in F on the second set of n variables by any 1-1 replacement of the variables in the first set by the variables in the second set. We likewise see that a closed F which does consist wholly of first order functions

---

[21]   Or iteratively closed system, or closed system.

possesses at least one function of each single variable in V;
and the totality of functions in F on one variable is transformed in 1-1 fashion into the totality of functions in F on
any other variable by replacing the first variable by the second.

We may therefore say that the functions in F on any one
set of n variables are merely repeated over all other sets of
n variables belonging to V. This redundancy can be overcome
as follows. Let V be simply ordered in the series $x_1, x_2, x_3,$.
...., and let $\mathfrak{F}$ be composed of those functions in F whose
arguments are $(x_1), (x_1, x_2), (x_1, x_2, x_3),$ etc. Then $\mathfrak{F}$ posesses
all of the essentially different functions to be found in F
without the latter's duplications.[22] $\mathfrak{F}$ will be called the
contracted closed system corresponding to F. Note that when F
consists wholly of first order functions, $\mathfrak{F}$ possesses only functions of x_1; otherwise, $\mathfrak{F}$ possesses at least one function of
$(x_1, x_2, \ldots x_n)$ for every positive integral n. We shall often
implicitly use the concept of a contracted closed system for the
purpose of exhibiting the functions possessed by a closed system.
On the other hand, the replacement process, which is the basis of
our concept of iterative closedness, demands the very duplication
of the 'same functions' over different sets of variables found
in closed systems, and deleted therefrom in forming contracted
closed systems. Our derivations will therefore explicitly refer
to closed systems as originally defined.[23]

---

[22]   Ultimate economy would be achieved by forming the set of
all 'abstract functions' corresponding to functions in F, two
functions being said to correspond to the same abstract function if they can be made identical by a 1-1 replacement of arguments. The resulting 'abstract closed system' corresponding
to F is, however, not as practical as $\mathfrak{F}$. It would carry us
too far afield to relate these ideas to the concept $f(\dot{x}_1, \dot{x}_2, \ldots$
$\dot{x}_n)$ of Principia Mathematica.

[23]   The requirement that V be a denumerably infinite class is
in agreement with the usual form of a propositional calculus.
Note that were W any infinite class of variables, G an iteratively closed system over W, then, if each function in G is
on a finite number of arguments, a system F iteratively closed
with respect to a denumerably infinite class V could be constructed such that the abstract closed systems, in the sense of
the last footnote, corresponding to G and F are identical.
If then V be infinite, it may as well be denumerably infinite.
V finite, however, leads to something new. Note that if F is
iteratively closed with respect to V, then if V' is an existent subclass of V, F' the class of functions in F with ar-

We give an independent definition of the system of functions
generated by a given existent finite set of functions. Variables and values are to be in V and v as heretofore. There
is no loss of generality in writing the functions that are to
generate the system in the form $f(x_1, x_2, \ldots x_{n_i})$, $i = 1, 2, \ldots v$,
$n_i \neq 0$. We then define the system of functions <u>generated</u> by
these generators over the given denumerably infinite class of
variables V as the class of all functions that can be assigned
to the system by the use of the following criterion. If X_1,
$X_2, \ldots X_{n_i}$, $i = 1, 2, \ldots v$, represent variables in V or functions in the generated system, then the function represented by
$f_i(X_1, X_2, \ldots X_{n_i})$ also belongs to the generated system. We shall
say also that every function thus obtainable can be generated by
the given generators. Note that in the initial applications of
this inductive definition $X_1, X_2, \ldots X_{n_i}$ can only be variables
in V.

It follows from this definition that each function in the
generated system can be expressed in finite form in terms of the
generators of the system, and variables in V. These expressions,
or operations as they may be called, can be classified according
to their <u>rank</u>. It is convenient to refer to variables in V as
operations of rank zero. Then if the highest rank of the oper-
ations $\mathfrak{X}_1, \mathfrak{X}_2, \ldots \mathfrak{X}_{n_i}$ is ρ, the operation $f_i(\mathfrak{X}_1, \mathfrak{X}_2, \ldots \mathfrak{X}_{n_i})$
will be said to be of rank $\rho+1$. We also inductively define
the <u>components</u> of $f_i(\mathfrak{X}_1, \mathfrak{X}_2, \ldots \mathfrak{X}_{n_i})$ to be $\mathfrak{X}_1, \mathfrak{X}_2, \ldots \mathfrak{X}_{n_i}$ and
their components, a variable in V being understood to have no
components. Clearly, an operation of rank ρ has at least one
component of each rank less than ρ. Thus, with $f_1(x_1, x_2)$ and
$f_2(x_1, x_2)$ as generators, $f_1\{f_2(x_1, x_2), f_1[f_2(x_3, x_1), x_2]\}$ is an
operation of rank three; and among its components are the operations $f_1[f_2(x_3, x_1), x_2], f_2(x_3, x_1), x_3$ of ranks two, one, zero,
respectively. Associated with the generated system of functions
is thus a system of operations. Each operation of rank greater
than zero in the latter system defines a unique function in the
former. On the other hand, for most of the generated systems

guments in V', then F' is iteratively closed with respect to
V'. This result easily transforms the solution of the principal
problem of the present paper, given for infinite V, into solutions of the same problem for finite V's. In this connection
see also the next footnote.

that we shall study, each function in the former system is yielded by an infinite number of different operations in the latter.

By means of these operations we readily verify that if $f(x_{i_1}, x_{i_2}, \ldots x_{i_n})$ is a function in a generated system on distinct variables $x_{i_1}, x_{i_2}, \ldots x_{i_n}$, and $X_1, X_2, \ldots X_n$ represent variables in V or functions in the system, then the function represented by $f(X_1, X_2, \ldots X_n)$ also belongs to the system. **That is, <u>every generated system is iteratively closed</u>.** Clearly, if each function in a given set of generators belongs to a closed system F, then the system of functions generated by those generators is contained in F. Hence, if in addition, each function in F can be generated by those generators, the generated system is F. There is no a-priori reason, however, for an arbitrary closed system admitting of generation by a finite set of generators.

We define the <u>order</u> of a finite set of generators as the highest order of the several functions in the set. If, then, a closed system can be generated by a finite set of generators, we define the <u>order</u> of the system as the lowest order that a finite set of generators of the system can have. The phrase, 'closed system of finite order' is then equivalent to the phrase 'generated system' provided the number of generators is understood to be finite.[24] Since functions of one variable can only generate functions of one variable, it follows that a closed system of the first order consists wholly of functions of the first order. On the other hand, a closed system of finite order greater than one must possess at least one function of order greater than one, and hence possesses functions of every order.

---

[24] The concept of generated system, however, continues to be valid if an infinite number of generators is allowed. This less stringent concept leads to a correspondingly wider concept of 'closed system of finite order'. The following observation assumes this wider concept. Clearly, a closed system over V of finite order n is determined by its n-th and lower order functions. It follows that if V' consists of say the first n variables in V, then there is thus determined a 1-1 correspondence between the closed systems over V of order less than or equal to n and the closed systems over V' in the sense of the preceding footnote. See also footnote 39 (§10). Where, as in the problem of the present paper, there are but a finite number of functions on a given finite set of arguments, there is no difference between the two concepts of 'closed system of order n'.

Continuing to restrict our attention to functions whose arguments are in V and values in v, we define a condition imposed upon such functions to be iterative[25] if whenever $f(x_{1_1},$ $x_{1_2},...x_{1_n})$, on distinct variables $x_{1_1},x_{1_2},...x_{1_n}$, satisfies the condition, and $X_1,X_2,...X_n$ represent variables in V, or functions satisfying the condition, then the function represented by $f(X_1,X_2,...X_n)$ also satisfies the condition.

Clearly, if each of several conditions is iterative, then the combined condition that a function satisfies each of those conditions is iterative.

There are two important consequences of a condition being iterative. On the one hand, the system of all functions satisfying the condition is iteratively closed. On the other, if each function in a given set of generators satisfies the condition, then every function in the generated system satisfies the condition. We shall in every case say that if each function in a closed system satisfies a certain iterative condition, then the system satisfies that condition. The discovery of iterative conditions thus plays a very important rôle in the determination of iteratively closed, and generated, systems of functions.

§2

TWO-VALUED SYSTEMS OF FUNCTIONS
and
THE RELATED LOGIC OF CLASSES

Variables, functions, and systems of functions for which the class of values v consists of but two members will be said to be two-valued. The following discussion is restricted to a single class v whose two members are symbolized + and - respectively.

Our definition of the identity of functions on the same set of variables implies the adoption of the extensional point of view with respect to functions. Every two-valued function,

---

[25]   We have borrowed the word "iterative" in these connections from Wiener (see footnote 6). But our "iterative condition" is not the same concept as Wiener's "iterative characteristic" [41].

$\phi(\beta,\delta,\ldots\eta)$ can therefore be uniformly exhibited in finite form by a table such as

$\beta\delta\ldots\eta$	$\phi(\beta,\delta,\ldots\eta)$
$++\ldots+$	$+$
$++\ldots-$	$-$
$\cdot\;\cdot\cdot\cdot\cdot\cdot$	$\cdot$
$\cdot\cdot\cdot\cdot\cdot\cdot$	$\cdot$
$\cdot\cdot\cdot\cdot\cdot\cdot$	$\cdot$
$--\ldots-$	$+$

which associates with every possible assignment of values to the variables $\beta,\delta,\ldots\eta$ the corresponding value of the function. Conversely, every such table defines a two-valued function of $\beta,\delta,\ldots\eta$. Since the two values $+$, $-$, can be assigned to n variables $\beta,\delta,\ldots\eta$ in 2^n different ways, it follows that there are exactly 2^{2^n} different two-valued functions of those n variables.

Let V be a given denumerably infinite set of variables $\alpha,\beta,\gamma,\ldots$. For each set of n variables belonging to V, with $n = 1,2,3\ldots$, construct the 2^{2^n} two-valued functions of those variables. The system composed of all of these two-valued functions will be called the <u>complete two-valued system over V</u>.

Clearly every iterative process built up out of two-valued variables and functions yields a two-valued function. It follows that the complete two-valued system over V is iteratively closed with respect to V.

The main problem of the present paper is to determine all of the closed two-valued systems over V. Since each of these closed systems is composed of functions belonging to the complete system, our problem can be restated as follows: to determine all of the closed subsystems over V of the complete two-valued system over V.

These systems are more easily visualized through their associated contracted systems. The 'complete contracted two-valued system over V' possesses exactly 2^{2^n} functions of order n, i.e., it is composed of 4 functions of the first order, 16 of the

second order, 256 of the third order, and so on. A contracted
closed two-valued system over V consisting wholly of first
order functions will be a subset of the set of four first order
functions in the complete contracted system. Every other contracted closed two-valued system over V, on the other hand,
will possess at least one of the 4 first order functions, at
least one of the 16 second order functions, at least one of the
256 third order functions, and so on, that make up the complete
contracted system. It may therefore be said to cut out an infinite swathe of functions from the complete contracted system.[26]

For reasons of presentation we replace the above abstract
problem by an equivalent problem in the logic of classes. Let
v' be the class of subclasses of a given existent class.[27] v'
then has at least two members, the given existent class, which
will be called the universal class and symbolized 1, and the
null class, symbolized 0. Let B,D,...H be a set of n independent variables whose admissible values are arbitrary classes
in v', and let f(B,D,...H) be a function of those variables
with its values also in v'. Among such functions we shall then
only be interested in those functions f(B,D,...H) below termed
membership functions.

Let x represent an arbitrary member of the universal class,
G a class belonging to v'. We define the <u>membership value of</u>
<u>x with respect to G</u> to be + when x is a member of G, -
when x is not a member of G. Each admissible assignment of
values to the variables x,B,D,...H thus determines the member-
ship values of x with respect to B,D,...H, and f(B,D,...H).
We then define f(B,D,...H) to be a <u>membership function</u> of B,
D,...H if the membership value of x with respect to f(B,D,..
..H) is found to depend solely on the membership values of x

---

[26] Actually these contracted systems are best visualized as
the different tables of the functions in the corresponding two-
valued systems when the headings of those tables are removed.

[27] For simplicity. It of course suffices for v' to be a
class of subclasses of the given existent class closed with respect to the operations logical sum, and complement with respect
to the given class.

with respect to B,D,...H.[28] Now each of the 2^n ways in which
membership values can formally be assigned to x with respect
to the n variables B,D,...H will be exhibited by actual values of x,B,D,...H. For we can let x be any member of the
universal class, and B,D,...H the universal or null class, according as the corresponding membership value is desired to be
+ or -. It follows that every membership function f(B,D,...H)
determines the <u>membership table</u> of the function, e.g.,

B D...H	f(B,D,...H)
+ +...+	+
+ +...-	-
. .....	.
. .....	.
. .....	.
- -...-	+

which gives the membership value of x with respect to f(B,D,.
...H) for each of the 2^n ways in which x may assume membership values with respect to B,D,...H.[29] Conversely, every membership table on B,D,...H determines a membership function
f(B,D,...H) of which it is the membership table. For, given
B,D,...H, such a table tells of any x in the universal class
that, according as it does or does not belong to the several
classes B,D,...H, it does or does not belong to the class,
f(B,D,...H), and thus determines the class f(B,D,...H). A 1-1
correspondence has thus been established between the different
membership functions of B,D,...H, and the membership tables on
B,D,...H. Evidently there are 2^{2^n} different membership tables
on B,D,...H. It follows that there are exactly 2^{2^n} different
membership functions of the n variables B,D,...H.

This suggests a close relationship between two-valued functions and membership functions, which may be formulated precise-

---

[28]   We cannot say 'Boolean' since the latter may involve in
its definition constants in v' other than 1 and 0.

[29]   Note that if the given universal class has at least 2^n
members, then a single suitable choice of B,D,...H can be
made with respect to which x will assume all of the 2^n possible sets of membership values as it roams through the universal class. The one corresponding class f(B,D,...H) suffices,
then, to determine the membership function f(B,D,...H). Hence,
in part, the efficacy of the Venn diagrams.

ly as follows. Let n two-valued variables $\beta, \delta, \dots \eta$ be set
into 1-1 correspondence with the n membership variables B,
D,...H. Then, if in the table of a two-valued function $\phi(\beta,$
$\delta, \dots \eta)$ the variables $\beta, \delta, \dots \eta$ are replaced by the corresponding variables B,D,...H, the result can be interpreted as
the membership table of a membership function f(B,D,...H), and
conversely. Through their tables a 1-1 correspondence is thus
set up between the 2^{2^n} two-valued functions of $\beta, \delta, \dots \eta$, and
the 2^{2^n} membership functions of B,D,...H.

As in the case of two-valued systems, we let V' be a given
denumerably infinite set of variables A,B,C,..., and define
the complete membership system over V' as the system composed
of all membership functions whose arguments, finite in number
for each function, belong to V'. Here, as below, it is understood that the variables and membership functions in question
all correspond to the above class of values v'. That the complete membership system over V' is iteratively closed with
respect to V' follows from the more general result proved
below. We then propose the problem of determining all iteratively closed 'membership systems' over V', and observe that this
is equivalent to determining all of the closed subsystems over
V' of the complete membership system over V'.

Now set the denumerably infinite set of variables $\alpha, \beta, \gamma, \dots$
forming V into 1-1 correspondence with the denumerably infin-
ite set of variables A,B,C,... forming V'. By setting up the
above correspondence between two-valued functions and membership
functions for each finite set of variables in V, and finite
set of corresponding variables in V', a 1-1 correspondence is
established between the functions in the complete two-valued
system over V and the functions in the complete membership
system over V'. This correspondence furthermore is preserved
under iteration. For let the iterative process $\phi(\mu, \nu, \dots \omega)$,
depending on variables $\beta, \delta, \dots \eta$ in V, correspond to the
iterative process f(M,N,...W), depending on the corresponding
variables B,D,...H in V', i.e., let the two iterative processes be similarly built up out of variables and functions that
correspond. Then, if membership values are assigned to x with
respect to B,D,...H which are respectively the same as values
assigned to $\beta, \delta, \dots \eta$, membership values of x with respect to

M,N,...W will be determined which are respectively the same as
the resulting values of μ,ν,...ω, and hence a membership value
of x with respect to f(M,N,...W) will be determined which is
the same as the value φ(μ,ν,...ω). The iterative process f(M,
N,...W) therefore yields a membership function of B,D,...H
which corresponds to the two-valued function of β,δ,...η yielded by the corresponding iterative process φ(μ,ν,...ω).

   Given then any closed two-valued system over V, the membership functions corresponding to the two-valued functions
therein must constitute a closed membership system over V',
and conversely. A 1-1 correspondence is thus induced between
the aggregate of all closed two-valued systems over V and the
aggregate of all closed membership systems over V'. In terms
of this correspondence a solution of the problem of determining
all closed two-valued systems over V automatically constitutes
a solution of the problem of determining all closed membership
systems over V', and conversely. As the problem for membership
systems will allow us to use the notation and algebra of symbolic
logic, we shall confine our attention to its solution, and consider the result a solution of the problem for two-valued systems.[30]

§3

THE JEVONS NOTATION; EXPANSIONS OF MEMBERSHIP FUNCTIONS

   Since our mathematical formulation of membership function
does not quite see eye to eye with the usual developments of
symbolic logic, we go to some length to bridge the resulting
gap. Our concern is not with formal symbolic logic, but with
its interpretation as a logic of classes. In this discipline
we may be concerned with the formal structure of an _expression_

---

[30]   Indeed our formulas for the functions in the various
closed membership systems, when considered merely in connection with the about to be defined complete expansions of those
functions, can be immediately translated into conditions on the
membership tables of those functions, and hence automatically
into conditions on the tables of the functions in the various
closed two-valued systems. This is illustrated in footnote 37,
(§8).

f(A,B,...I), or, when this expression is meaningfully interpreted, we may refer to the _function_ f(A,B,...I), and, when A, B,...I are assumed to take on certain classes in v' as values, our concern may then be with the corresponding _class_ f(A,B,....I), also in v'. If, for example, f(A,B,...I) is the null class o for every assignment to A,B,...I of values in v', we shall say that f(A,B,...I) is identically o, and write f(A,B,...I) = o. Our use of the word 'contain' will always refer to inclusion of classes represented, and not to the formal structure of corresponding representations. Exactly, if f(A,B,...I) and g(A,B,...I), with not all arguments necessarily present in each expression, are such that for each assignment of values in v' to A,B,...I the class f(A,B,...I) contains the class g(A,B,...I), then we shall say that the function f(A,B,...I), or even the expression f(A,B,...I), contains the function, or expression, g(A,B,...I). If, on the other hand, for each assignment of values in v' to A,B,...I the classes f(A,B,...I) and g(A,B,...I) have no members in common, we shall say that f(A,B,...I) excludes g(A,B,...I).[31]

We define the negative of an arbitrary class belonging to v' as the class which consists of those members of the universal class which are not members of the given class. With logical sum and logical product defined as usual, it follows that the logical sum of a class and its negative is always the universal class 1, the logical product, the null class o. Furthermore, the negative of 1 is o, of o, 1.

Following Jevons we introduce the small letters a,b,c,... to represent the negatives of the classes represented by the corresponding capital letters A,B,C,... . By a _complete term_ on n given letters A,B,...I we shall mean the formal logical product of the letters belonging to any selection chosen from the n pairs of letters (A,a), (B,b), ... (I,i). Thus AbcDe is a complete term on the five letters A,B,C,D,E. For the n letters A,B,...I there are 2^n such selections, and hence, provided we disregard the particular order of their 'factors', there are 2^n complete terms on A,B,...I. They constitute the

---

[31] The paragraph just concluded is a recent addition. Likewise a few minor changes in the rest of this section.

logical alphabet of Jevons for those letters.[32]

If t(A,B,...I) represents a complete term on A,B,...I,
then each assignment of classes in v' as values of A,B,...I
results in a value of t(A,B,...I) which is also a class in v'.
It is then significant that <u>t(A,B,...I) is never 0 for all</u>
<u>values of A,B,...I</u>. In fact, by letting the variables A,B,...
..I assume the values 1 or 0 according as the corresponding letters are capital or small in t(A,B,...I), t(A,B,...I)
assumes the value 1. The relationship between a class and its
negative yields the formal identities A+a =1, Aa = 0. Since
for any two distinct complete terms on A,B,...I there is at
least one letter that is capital in one term and small in the
other, it follows that:

> The logical product of any two distinct complete
> terms on A,B,...I is identically 0. On the other hand,
> by expanding the first member of the identity (A+a)(B+b)
> ...(I+i) = 1, we see that the logical sum of all the
> complete terms on A,B,...I is identically 1.

We have, of course, assumed here the ordinary formal laws
of logical sum and logical product. Likewise in later developments.

Given a membership function f(A,B,...I), if we represent
the class f(A,B,...I) by M, the negative of f(A,B,...I)
may be represented by m. The membership table of f(A,B,...I)
can then be rewritten in the form

A B...I	M
A B...i	m
. .....	.
. .....	.
. .....	.
a b...i	M

where the last column assigns an arbitrary member of the univer-
sal class to M or m according as it belongs to A or a,
B or b, ... I or i. From this table we see that if t(A,B,.
...I) is a complete term on A,B,...I, then either the class
t(A,B,...I) is contained in the class M for all values of
A,B,...I, or the class t(A,B,...I) is contained in the class

---

[32]   See, for example, [12].

m for all values of A,B,...I. That is, we have identically in
A,B,...I either t(A,B,...I)M = t(A,B,...I) or t(A,B,...I)m
= t(A,B,...I). Since Mm = 0, we also have in the first case
t(A,B,...I)m = 0, in the second t(A,B,...I)M = 0.

Hence the complete term t(A,B,...I) is either
contained in M and excluded from m, or contained
in m and excluded from M.

As t(A,B,...I) cannot be identically 0, these two alter-
natives are mutually exclusive. Now let ε_1(A,B,...I) be the
formal logical sum of the complete terms on A,B,...I contained
in M, ε_2(A,B,...I) of those contained in m. By multiplying
both sides of the identity AB...I+AB...i+...+ab...i = 1 by M,
and again by m, we see that ε_1(A,B,...I) is identically
equal to M, ε_2(A,B,...I) to m.

The ordered pair of expressions ε_1(A,B,...I): ε_2(A,B,...I)
will be called the complete expansion of the membership function
f(A,B,...I). The first expression, ε_1(A,B,...I) of the complete expansion of f(A,B,...I) is then a formal representation
of f(A,B,...I), the second, ε_2(A,B,...I), of the negative
of f(A,B,...I). If, on the other hand, we do not start with a
given membership function of A,B,...I, but arbitrarily form
a pair of expressions ε_1(A,B,...I): ε_2(A,B,...I) by assign-
ing each of the 2^n complete terms on A,B,...I to one or the
other of these expressions, the result will be called a complete
expansion on the n letters A,B,...I. Thus ABC+ABc+AbC+aBC:
abc+abC+aBc+Abc is a complete expansion on the three letters,
A,B,C. Clearly there are exactly 2^{2^n} complete expansions on
the n letters A,B,...I provided we disregard the particular
order of the terms within each expression of a complete expansion. With this proviso, we see that each membership function
of A,B,...I determines its complete expansion. Conversely,
each complete expansion on A,B,...I determines a membership
table on A,B,...I, and hence a membership function of A,B,..
...I of which it is the complete expansion. A 1-1 correspondence is thus set up between the 2^{2^n} membership functions of
A,B,...I, and the 2^{2^n} complete expansions on A,B,...I.

A complete term on any proper subset of a given set of let-

ters A,B,...I will be called an incomplete term on A,B,...I.
Thus, bDe is an incomplete term on the five letters A,B,C,D,
E. Clearly, an incomplete term on A,B,...I also assumes a
value in v' for each assignment of values in v! to A,B,...
..I. Given an incomplete term, and any complete term, on A,B,.
...I, then either each of the letters present in the incomplete
term is capital in both terms or small in both terms, or at
least one letter is capital in one term and small in the other.
In the first case the logical product of the two terms is the
complete term, in the second the logical product is identically
O.

> That is, any complete term on A,B,...I is either
> contained in a given incomplete term on A,B,...I or
> is excluded therefrom.

Clearly, every complete term on A,B,...I contained in a
given incomplete term on A,B,...I is the logical product of
the incomplete term and a complete term on those letters A,B,..
...I which are missing from the incomplete term, and conversely. Since the logical sum of all the complete terms on those
missing letters is identically 1, we see by multiplying this sum
by the incomplete term that:

> An incomplete term on A,B,...I is identically
> equal to the logical sum of all the complete terms
> on A,B,...I contained therein.

An ordered pair of expressions such that each is the formal
logical sum of terms, complete or incomplete, on A,B,...I, and
such that either expression is identically equal to the negative
of the other, will be called an expansion on the letters A,B,..
...I. Given a membership function of A,B,...I, f(A,B,...I),
then an expansion on A,B,...I will be called an expansion of
the membership function provided the first expression of the
expansion is identically equal to f(A,B,...I), and hence the
second to the negative of f(A,B,...I). Thus, the logical sum
of A and B, which is a membership function of A and B
with the complete expansion AB+Ab+aB: ab is easily seen to

admit the other expansions A+aB:ab, Ab+B:ab, A+B:ab.

The condition that one expression of an expansion on A,B,.
...I is identically equal to the negative of the other is equivalent to the pair of conditions: the logical sum of the two
expressions is identically 1, the logical product identically
0.

It follows that every complete term on A,B,...I
is contained in at least one term of an expansion on
A,B,...I; as otherwise it would be excluded from every
term of the expansion, and the logical sum of the two
expressions of the expansion would not be 1. Furthermore, a complete term on A,B,...I cannot be contained
in terms belonging to different expressions of an expansion on A,B,...I, or the logical product of the
two expressions would not be 0.

If then each term of an expansion on A,B,...I is replaced
by the logical sum of all the complete terms on A,B,...I contained therein, and identical terms are then united, the result
will be a complete expansion on A,B,...I whose first and second expressions are identically equal to the first and second
expressions respectively of the given expansion. Since every
complete expansion on A,B,...I defines a membership function
of A,B,...I, we thus see that every expansion on A,B,...I is
an expansion of a membership function of A,B,...I.

In the above general discussion there occur certain lacunae
which we now fill in. We have described the expressions of both
complete expansions, and expansions in general, to be formal
logical sums of terms, whereas the number of terms to be assigned
to an expression may be one or zero. In the first case the expression in question must clearly be the term itself. In the
second case we can consistently write 0 for the expression.
For the complete expansion then has all of its terms in the
other expression, and this other expression is therefore identically equal to 1. Note that 0 must not be considered a term
of the expansion. On the other hand, it is convenient nominally
to consider 1 a complete term on no letters, and hence actually
an incomplete term on every set of letters. We can then state

that a function whose complete expansion has 0 for one expression admits an expansion whose other expression is 1.

§4

DIFFERENT EXPANSIONS OF THE SAME FUNCTION;

RELEVANT AND IRRELEVANT VARIABLES

We saw in the preceding section that every complete term
on A,B,...I is contained in at least one term of an arbitrary
expansion on A,B,...I, and that the terms of this expansion
containing a given complete term on A,B,...I are all in the
same expression of the expansion. We now readily see that the
terms of different expansions of the same membership function
f(A,B,...I) that contain a given complete term on A,B,...I
are in the first expression for all expansions of the function,
or in the second expression for all expansions of the function.
For otherwise the complete term would be contained both in f(A,
B,...I) and in the negative of f(A,B,...I).

We may then say that the terms of different expansions of the same membership function f(A,B,...I)
that contain a given complete term on A,B,...I are in
corresponding expressions of their respective expansions.
It follows, in particular, that every term of either
expression of the complete expansion of f(A,B,...I)
is contained in at least one term of the corresponding
expression of any other expansion of f(A,B,...I).

If a membership function f(A,B,...I) admits an expansion
which does not involve an argument G of the function, G will
be said to be an irrelevant variable of the function; otherwise
relevant.[33]

---

[33]   Note that it is the membership table of a membership function that determines for us what the arguments of the function
are. If, then, a membership function is given by an expansion
thereof, we assume that it is explicitly stated what other variables, if any, than those which appear in the expansion are arguments of the function.

Every expansion of f(A,B,...I) therefore involves all the relevant variables of f(A,Ḃ,...I).
We now prove that f(A,B,...I) admits an expansion
involving only its relevant variables. In particular,
if all the arguments of f(A,B,...I) are irrelevant,
f(A,B,...I) must admit one of the two expansions
1:0, 0:1.

If G is an irrelevant variable of f(A,B,...I), then any
two complete terms on A,B,...I which differ only in the capitalization of G must be in the same expression of the complete
expansion of f(A,B,...I). For in any expansion of f(A,B,...I)
which does not involve G, the same term which contains one of
such a pair of complete terms must contain the other. By successive application of this result it follows that any two complete terms on A,B,...I which differ only in the capitalization of irrelevant variables of f(A,B,...I) will be in the
same expression of the complete expansion of f(A,B,...I). Let
B,D,...H be the relevant variables of f(A,B,...I). The complete terms on A,B,...I can be grouped into mutually exclusive
sets of terms, each set consisting of the complete terms on A,
B,...I contained in a given complete term on B,D,...H. Since
the terms in any one of these sets differ only in the capitalization of irrelevant variables of f(A,B,...I), they will all
be in the same expression of the complete expansion of f(A,B,
...I). Furthermore, their logical sum is identically equal to
the complete term on B,D,...H in which they are all contained.
By thus uniting the terms in each of the above sets, the complete
expansion of f(A,B,...I) is transformed into an expansion of
f(A,B,...I) involving only the variables B,D,...H.

A special consequence of the preceding argument is the following criterion for determining the relevant and irrelevant
variables of f(A,B,...I). The necessary and sufficient condition that an argument G of a membership function f(A,B,...I)
be irrelevant is that the complete expansion of f(A,B,...I)
remain unchanged when G and g are interchanged, and hence:

That the function f(A,B,...I) remain unchanged when

the variable G is replaced by its negative.[34]

ITERATIVE ASPECTS OF THE JEVONS NOTATION

Let each function in a closed membership system over V'
be given by some expansion of that function. In terms of these
expansions the process of replacing the arguments of a function
in the system by variables in V' or functions in the system
assumes the following form.

In an expansion of the given function replace the
capital and small letter of each argument of the function by the capital and small letter respectively of a
replacing variable, or by the first and second expressions respectively of an expansion of a replacing function. The result, after due simplification by the rules
of symbolic logic, will be an expansion of the function
yielded by the given iterative process.

If an argument G of a function is thus to be replaced by
a variable E or by a function with expansion $\varepsilon_1:\varepsilon_2$ we shall
say that in the expansion of the function, G:g is to be re-
placed by E:e or $\varepsilon_1:\varepsilon_2$ respectively, i.e., G is to be replaced by E or ε_1 , g by e or ε_2 .

Our notation f(M,N,...W) for an iterative process depending on variables A,B,...I does not reveal the identity of the
original arguments of f. It is convenient in their place to
use the auxiliary symbols M,N,...W which represent the variables and functions that replace those arguments. Then, from an
expansion of f(M,N,...W) in terms of M,N,...W we obtain by
the above method an expansion of f(M,N,...W) in terms of A,
B,...I, that is, an expansion of the function of A,B,...I

---

[34]   We have endeavored to develop a technique expressed entirely in terms of expansions of membership functions. Otherwise
we would have defined an argument G of f(A,B,...I) to be
an irrelevant variable thereof if the value of f(A,B,...I) is
independent of the value of G. This is easily restated in
terms of the membership table of f(A,B,...I).

yielded by the iterative process f(M,N,...W).

In particular we shall be interested in the relationship between the complete expansion of f(M,N,...W) in terms of M, N,...W and the complete expansion of f(M,N,...W) in terms of A,B,...I. To study this relationship we must first extend a result of §3. We saw there that if M and m represent a membership function of A,B,...I and its negative, then every complete term t(A,B,...I) is either contained in M and excluded from m, or contained in m and excluded from M. We now observe that this result continues to hold if M and m represent one of the variables A,B,...I and its negative, or a membership function of only some of the variables A,B,...I, and its negative. The first case is immediate. In the second, note that an expansion of the function can be considered to be an expansion on all the letters A,B,...I. Since t(A,B,...I) is therefore contained in some term of one expression of the expansion and excluded from every term of the other expression it will be contained in one expression of the expansion and excluded from the other. Whence the above result.

For complete generality let f(M,N,...W) depend on no other variables than A,B,...I. A complete term t(A,B,...I) will then be contained in one member of each pair (M,m),(N,n),....
(W,w) and be excluded from the other. It therefore determines a selection from these pairs such that it is contained in each member of that selection but is excluded from at least one member of any other selection from those pairs. t(A,B,...I) is therefore contained in the complete term on M,N,...W determined by this selection but is excluded from every other complete term on M,N,...W. Clearly, the complete term on M,N,..
...W containing t(A,B,...I) is in the first, or second, expression of the complete expansion of f(M,N,...W) with respect to M,N,...W according as t(A,B,...I) is contained in f(M,N, ...W), or in the negative of f(M,N,...W). It follows, in particular, that:

If f(M,N,...W) depends on all the variables A,B,...I, each term of either expression of the complete expansion of f(M,Ṅ,...W) in terms of A,B,...
...I is contained in one and only one term of the cor-

responding expression of the complete expansion of
f(M,N,...W) in terms of M,N,...W.

Finally let f(M,N,...W) and f(M',N',...W') be two iterative processes with the same f which depend on no other variables than A,B,...I. It will be convenient to refer to G:g
as an expansion of the variable G. By corresponding complete
terms on M,N,...W and M',N',...W' will be meant two complete
terms T(M,N,...W) and T(M',N',...W') with the same T. Re-
calling that a membership function and its negative are identically equal to the first and second expressions respectively of
any expansion of the function, we are led by the above argument
to the following result.

If a complete term t(A,B,...I) is contained
in corresponding expressions of expansions of M,N,.
...W and M',N',...W' respectively, it will be contained in corresponding complete terms on M,N,...W
and M',N',...W', and hence in corresponding expressions of expansions of f(M,N,...W) and f(M',N',..
...W').

§6

DUALITY[35]

Let us temporarily set aside the Jevons notation and symbolize the negative of A by -A. Since -(-A) = A it follows that the function 'negation' is its own inverse. We now
define the <u>dual</u> of a membership function f(A,B,...I) as the
transform of the function under negation. The dual of f(A,B,..

---

[35] The concept of duality goes back at least to Schröder. The
writer came upon this concept early in the solution of the main
problem of the present paper via the relativity of the signs, +,
-, occurring in a truth-table when they are considered to be just
marks. Zylinski [42] seems to have had the same experience. In
the present introduction of this concept we follow the formulation of Wiener [39] and Frink [8]. The results of the present
section in part overlap those of Bernstein in [3]. See also Frink
[8], p. 485.

...I) is then -f(-A,-B,...-I). Clearly, the dual of the dual
of f(A,B,...I) is f(A,B,...I). We can therefore refer to a
membership function and its dual as a pair of dual functions.

In connection with iteration it is convenient to call a
variable its own dual. If in an iterative process f(M,N,...W)
we replace f,M,N,...W by their duals, the resulting iterative
process will be termed the dual of the given process. It follows from the general property of transforms that dual iterative
processes yield dual functions.

Hence, if we replace each function in a closed
membership system by its dual, the resulting system
of functions will be closed. This closed system will
be called the dual of the given system. Clearly, the
dual of the dual of a closed system is the system itself. We can therefore talk of a pair of dual systems.

If each function in a set of generators is replaced by its dual,
the resulting set of generators will be called the dual of the
given set. Since dual iterative processes yield dual functions,
it follows by induction that <u>dual sets of generators generate
dual closed systems</u>. This principle introduces a great economy
in the study of closed membership systems.

A function, or closed system, which is its own dual will
be termed <u>self-dual</u>. Clearly, a closed system consisting wholly of self-dual functions is self-dual. On the other hand, a
self-dual system need not consist wholly of self-dual functions.
All we can say is that the dual of each function in a self-dual
system is also in the system.

The condition of self-duality as applied to membership functions is our first example of an iterative
condition.

In fact, if each function used in building up an iterative process is self-dual, then the iterative process will be its own
dual. Hence the function yielded by the process will be its own
dual, that is, self-dual. It follows in particular that the
closed system generated by a set of self-dual generators is a

self-dual system consisting wholly of self-dual functions.

When our definition of the dual of a function is
translated into the Jevons notation, we obtain the following rule for writing down an expansion of the dual
of a function given an expansion of the function. Replace each capital letter occurring in the expansion by
the corresponding small letter and conversely, and interchange the resulting expressions. The particular expansion yielded by this rule may be called the dual of
the given expansion.

Thus, the dual of the expansion A+B:ab is AB:a+b; and
hence the dual of the membership function 'logical sum of A
and B' is the membership function 'logical product of A and
B'.

If each capital letter occurring in a term is replaced by
the corresponding small letter, and conversely, the resulting
term will be called the dual of the given term. Thus the dual
of AbcDe is aBCdE. We can therefore restate the above rule
for obtaining the dual of a given expansion as follows. Place
the dual of each term of the given expansion in the opposite
expression of the desired dual expansion. Now the complete
terms on a given set of letters can be grouped into mutually
exclusive pairs of dual terms. Hence, the above process, when
applied to the complete expansion of a function, results in the
complete expansion of the dual of the function. A function will
therefore be self-dual when and only when its complete expansion
is unchanged by this process.

It follows that in the complete expansion of a
self-dual function dual terms are always in opposite
expressions, while in the complete expansion of a
function that is not self-dual at least one pair of
dual terms are in the same expression.

Two complete terms on the same set of letters are evidently
dual when and only when no letter is capital in both terms or
small in both terms. Hence a function is non-self-dual when and

only when there are two terms in the same expression of its complete expansion which have neither a capital nor small letter in common.[36] We now show that this criterion of non-self-duality applies to any expansion of a function, provided 1 is not a term of this expansion. In fact, if two terms in the same expression of the complete expansion of the function have neither a capital nor a small letter in common, two terms of the given expansion which respectively contain them, and which therefore are in the same expression of the given expansion, will have neither a capital nor a small letter in common. Conversely, if two terms in the same expression of the given expansion have neither a capital nor a small letter in common, by supplying missing letters in the two terms with opposite capitalizations two complete terms in the same expression of the complete expansion are obtained which have neither a capital nor a small letter in common.

We can therefore state that the necessary and sufficient condition that an expansion which does not have 1 for a term represent a self-dual function is that each pair of terms belonging to the same expression of the expansion have either a capital or a small letter in common.

§7

UNIFORM TERMS

A term will be said to be a uniform term of the first kind if all of its letters are capital, of the second kind if all of its letters are small.

1 is then the only term which is a uniform term of both kinds. A complete expansion on the letters A,B,...I has exactly one uniform term of the first kind and one of the second, namely, A,B,...I and a,b,...i. Since each term of the complete expansion of a function is contained in at least one term of any

---

[36] Note that 1:0 and 0:1 are never complete expansions.

other expansion of the function, while a term containing a uniform term of a given kind must be a uniform term of the same kind,

It follows that every expansion of a function has at least one uniform term of the first kind and at least one uniform term of the second kind; and if the complete expansion of the function has its uniform term of the i-th kind in the j-th expression, then every expansion of the function has all of its uniform terms of the i-th kind in the j-th expression.

The condition that the uniform terms of the first kind are in the first expression of an expansion of a function is iterative.

For let $f(M,N,...W)$ be an iterative process depending on the variables $A,B,...I$, and built up out of functions satisfying the condition. $AB...I$ will be contained in every uniform term of the first kind which involves no other letters than $A,B,...I$. It will therefore be contained in $M,N,...W$, and hence in $MN...W$. But $MN...W$ is in the first expression of the complete expansion of $f(M,N,...W)$ in terms of $M,N,...W$. Hence $AB...I$ will be in the first expression of the complete expansion of $f(M,N,...W)$ in terms of $A,B,...I$.

By the dual argument it follows that the condition that the uniform terms of the second kind are in the second expression of an expansion of a function is also iterative. Therefore, the combined condition, the uniform terms of the first kind are in the first expression, the uniform terms of the second kind are in the second expression of an expansion of a function, is iterative.

§8

THE [A:a], [AA:] AND [:aa] CONDITION

We consider now the following three analogous conditions
that an expansion of a function may satisfy.

The [A:a] condition: for any two terms in different expressions of the expansion there is a letter
which is capital in the term that is in the first expression and small in the term that is in the second
expression.
The [AA:] condition: any two terms of the first
expression of the expansion have a capital letter in
common.
The [:aa] condition: any two terms of the second
expression of the expansion have a small letter in
common.

In the last two conditions we must let the phrase 'two
terms' cover the case of a single term chosen twice. This qual-
ification is required only when the expression referred to in
the condition consists of but a single term, in which case the
condition is then equivalent to the term in question possessing
a letter with the indicated capitalization. Note that when an
expression referred to in any of the three conditions is 0,
that expression has no terms, and hence the condition in question must be considered to be satisfied. Thus 1:0 satisfies
both the [A:a] and [:aa] condition, 0:1 the [A:a] and
[AA:] condition.

We first prove that if one expansion of a function
satisfies any of these three conditions, then every expansion of the function satisfies that condition.

Since the proof is much the same for all three cases, we
give it explicitly only for the [A:a] condition. If the [A:a]
condition is not satisfied by the complete expansion of a function, there will be a term in the first expression of this com-

plete expansion, and a term in the second expression, such that
no letter is at the same time capital in the first term and
small in the second. Any other expansion of the same function
will have a term in its first expression and a term in its second expression which respectively contain these terms of the
complete expansion, and hence can be obtained from them by omitting some of their letters. There is therefore no letter which
is capital in the first of these containing terms and small in
the second. Hence the new expansion also fails to satisfy the
[A:a] condition. Conversely, if any expansion of a function
fails to satisfy the [A:a] condition, there will be a term in
the first expression, and a term in the second expression, of
the expansion which lead to this failure. And by supplying missing letters in the first term as small letters, and in the
second term as capital letters, two complete terms are obtained
which result in the complete expansion of the function failing
to satisfy the [A:a] condition. It follows that the [A:a]
condition is satisfied by all expansions of a function or by
none.

Since these conditions on an expansion of a function are
independent of the particular expansion of the function employed,
they may be considered to be conditions on the function itself.[37]

It is then significant that each of these conditions, considered as a condition imposed on a function, is iterative.

Again the proofs are similar so that we restrict our attention to a single condition, this time the [AA:] condition.
Let f(M,N,...W) be an iterative process depending on the variables A,B,...I and built up out of functions satisfying the

---

[37]   Thus the [:aa] condition may be restated: whenever both
x_1 and x_2 of the universal class are outside of the class
f(A,B,...I), then x_1 and x_2 are both outside of at least one
of the classes A,B,... I. In terms of the membership table of
f(A,B,...I) this becomes, whenever the membership value of x
with respect to f(A,B,...I) is - for each of two assignments
of membership values of x with respect to A,B,...I, then the
membership value of x with respect to at least one of the variables A,B,...I is - in each assignment. In our original
presentation in terms of truth-tables this condition was briefly
written, "any two - configurations have a - in common".

[AA:] condition. Let t_1 and t_2 be any two terms of the first expression of the complete expansion of $f(M,N,\dots W)$ in terms of $A,B,\dots I$. t_1 and t_2 will then be respectively contained in two terms T_1 and T_2 of the first expression of the complete expansion of $f(M,N,\dots W)$ in terms of $M,N,\dots W$. Since f satisfies the [AA:] condition, T_1 and T_2 must have a capital letter, say P, in common. Both t_1 and t_2 will therefore be contained in P. If P is one of the letters A, $B,\dots I$, t_1 and t_2 have that capital letter in common. Otherwise, t_1 and t_2 will be respectively contained in two terms t_1' and t_2' of the first expression of some expansion of the function P. As this function satisfies the [AA:] condition, t_1' and t_2' will have a capital letter in common, and hence t_1 and t_2 have that capital letter in common. The function of $A,B,\dots I$ yielded by the iterative process $f(M,N,\dots W)$ therefore satisfies the [AA:] condition.

The [A:a] condition assumes an added importance from the fact that a function satisfying this condition admits a characteristic form of expansion which is in general different from, and usually more useful than the complete expansion of the function.

We proceed to show that every function satisfying the [A:a] condition admits an expansion whose first expression involves no small letters and whose second expression involves no capital letters.

Consider any term of the first expression of the complete expansion of an [A:a] function $f(A,B,\dots I)$ which involves at least one small letter. Every term obtainable from this term by changing one or more of its small letters into the corresponding capital letters willlthen also be in that first expression. For such a term can have no letter small that is capital in the given term. Now these terms, together with the given term, constitute all of the complete terms on $A,B,\dots I$ that are contained in the incomplete term on $A,B,\dots I$ formed from the given term by omitting all of its small letters. This incomplete term on $A,B,\dots I$, which contains the given term, and involves no small letters, is therefore itself contained in the first ex-

pression of the complete expansion of f(A,B,...I). We can therefore replace every term of that first expression which involves
at least one small letter by its associated incomplete term and
thereby transform that first expression into one involving no
small letters. By the dual method, the second expression of the
complete expansion of f(A,B,...I) can be transformed into one
which involves no capital letters.

Conversely, if a function admits an expansion whose first
expression involves no small letters, and whose second expression involves no capital letters, the function must satisfy the
[A:a] condition. In fact, this must be true if either expression is of the indicated form. For in any expansion, corresponding to a term in the first expression and a term in the second expression there must be a letter that is capital in one
term and small in the other; as otherwise the logical product
of the two expressions would not be identically 0. And if either
expression is of the indicated form, it must be the term in the
first expression that involves the capital letter, the term in
the second expression that involves the small letter. It follows that if either expression of an expansion of a function is
of the indicated form, the function admits an expansion in which
both expressions are of that form.

If in such an expansion of an [A:a] function identical
terms are united, and each term which is contained in another
term of this simplified expansion is then removed, the result
will be an expansion of the function, of the same form as the
given expansion, in which no term is contained in another term.
This expansion will be called the normal expansion of the [A:a]
function. It can be proved that an [A:a] function admits but
one normal expansion.[38] Furthermore, every variable appearing
in the normal expansion of an [A:a] function is relevant.
Clearly, when one expression of the complete expansion of an
[A:a] function is 0, the normal expansion of the function is
either 1:0 or 0:1. In every other case each expression of
the normal expansion of an [A:a] function has at least one
term, and the first expression is actually written in terms of
capital letters only, the second in terms of small letters only.

---

[38]   A paper of Blake's [6] may here be relevant.

As expamples we have A+B:ab and AB:a+b as the normal expansions of the [A:a] functions 'logical sum of A and B' and
'logical product of A and B' respectively.

The three conditions discussed in the present section are
intimately associated with the condition of self-duality. Note
in general that if a function satisfies the [A:a] condition,
so will the dual of the function. And if a function satisfies
one of the two conditions [AA:], [:aa] the dual of the function will satisfy the other. The [A:a] condition may therefore be thought of as a self-dual condition, the [AA:] and
[:aa] conditions as duals of each other. It follows immediately
that if a self-dual function satisfies one of the two conditions
[AA:], [:aa], it also satisfies the other. Furthermore, a self-
dual function which satisfies one of the two conditions [A:a],
[AA:] also satisfies the other. For let t_1 be any term in the
first expression of the complete expansion of a self-dual function, t_2 any term in the second expression. t_2', the dual of
t_2, will therefore be in the first expression. If t_2 has no
small letter which is capital in t_1, t_2' will have no capital
letter in common with t_1, and conversely. That is, if a self-
dual function fails to satisfy the [A:a] condition it also
fails to satisfy the [AA:] condition, and conversely. We therefore have that a self-dual function which satisfies any one of
the three conditions [A:a], [AA:], [:aa] satisfies the other
two as well. Finally, if a function satisfies both the [AA:]
and [:aa] condition, an expansion of the function cannot have
1 for a term, while any two terms in the same expression of the
expansion will have either a capital or a small letter in common.
The function will therefore be self-dual.

Hence, except for the pairs of conditions [A:a],
[AA:], and [A:a], [:aa], if a function satisfies
two of the four conditions, self-duality, [A:a], [AA:],
[:aa], it will also satisfy the other two.

In the development of Part II the failure of a function to
satisfy one of the three conditions studied in the present section plays a significant rôle. Such failure is always due to
the existence of a pair of terms in suitable expressions of an

expansion of the function which do not conform to the require-
ments of the condition. Now for any two terms of a complete expansion, the letters on which the expansion is written fall into
at most four groups with respect to their capitalization in the
terms in question, to wit,

> first term: capital capital small small,
> second term: small   capital small capital.

In testing the complete expansion for the [A:a] condition, the
first term will be understood to be in the first expression, the
second in the second expression; while for the [AA:] condition
both terms are to be in the first expression, for the [:aa] condition both terms are to be in the second expression. Then the
failure of the complete expansion to satisfy the [A:a], [AA:],
or [:aa] condition is equivalent to the existence of a pair of
terms in which no letter belongs to the first group, second group
and third group respectively.

§9

A AND a-COMPONENTS

The 2^n complete terms on n letters A,B,...I can be obtained by logically multiplying each of the 2^{n-1} complete terms
on the (n-1) letters B,...I by A, and again by a. If then
in a complete expansion on A,B,...I only terms involving the
capital letter A are retained, and that capital letter is then
struck out, the result will be a complete expansion on B,...I.
The function of B,...I defined by this complete expansion will
be called the A-component of the function f(A,B,...I) defined
by the original complete expansion. In like manner we define the
a-component of f(A,B,...I). For an arbitrary expansion of f(A,
B,...I) the corresponding rules are easily seen to be the following.

To obtain an expansion of the A-component of
f(A,B,...I) strike out capital A and all terms involving small a. Likewise for the a-component.

Thus AB+BC+AC:ab+bc+ac is an expansion with A-component
B+BC+C:bc, a-component BC:b+bc+c, i.e., B+C:bc, BC:b+c respectively.

If we symbolize the A and a-components of f(A,B,...I)
by $f_A(B,...I)$ and $f_a(B,...I)$ respectively, we see from their
definition that

$$f(A,B,...I) = Af_A(B,...I)+af_a(B,...I).$$

Now in this identity substitute 1 for A, and hence 0 for a,
and again 0 for ·A, and hence 1 for a. We then obtain the
formulas

$$f_A(B,...I) = f(1,B,...I),$$

$$f_a(B,...I) = f(0,B,...I).$$

Our formula for f(A,B,...I) shows that f(A,B,...I) is
determined by its A and a-components. Now a condition κ imposed upon functions may have the property that no two distinct
κ-functions can have the same A-component. We may then say that
a κ-function is determined by its A-component. Under certain
circumstances, as when κ is iterative, this property of κ enables us to determine the system of functions generated by a set
of κ-functions when the system of functions generated by the A-
components of the given functions is known. Though the details
of this method are best considered in connection with individual
cases as they come up in Part II, we prove here a theorem which,
with certain analogous results, constitutes the backbone of the
method.

THEOREM. Let $f_1(A,B,C)$ and $f_2(A,B,C)$ have A-components $\phi_1(B,C)$ and $\phi_2(B,C)$ respectively. Then
each function $\phi(B,C,...I)$ generated by ϕ_1 and ϕ_2
is the A-component of a function f(A,B,C,...I) generated by f_1 and f_2.

For consider any operation built up by ϕ_1 and ϕ_2 which
yields the function $\phi(B,C,...I)$. Since $\phi_1(B,C) = f_1(1,B,C)$,

and $\phi_2(B,C) = f_2(1,B,C,)$ we can rewrite this operation so that it is expressed in terms of $f_1, f_2, B, C, \ldots I$ and the constant 1. In this transformed operation replace each 1 by A. The new operation then yields a function $f(A,B,C,\ldots I)$ generated by f_1 and f_2. From the manner in which this new operation was obtained we see then that $\phi(B,C,\ldots I) = f(1,B,C,\ldots I)$. That is $\phi(B,C,\ldots I)$ is the A-component of $f(A,B,C,\ldots I)$.

Clearly the above theorem is true if in place of $f_1(A,B,C)$ with A-component $\phi_1(B,C)$ we have a function $f_1(A,B)$ with A-component $\phi_1(B)$. Furthermore completely analogous results hold for a-components. We shall refer to the theorem of this section when any of these results are required. In certain cases we shall also require the following observation which is a direct consequence of the way in which the above operation for $f(A,B, C,\ldots I)$ was constructed.

To wit, $f(A,B,C,\ldots I)$ can be generated by f_1 and f_2 through iterative processes in which no replacement is effected upon the variable A.

PART II

DERIVATION OF CLOSED SYSTEMS

§10

FIRST ORDER FUNCTIONS AND SYSTEMS AND THEIR GENERAL SIGNIFICANCE

There are four complete expansions on a single variable A, to wit, A:a, A+a:0, 0:A+a, a:A. The corresponding functions of A will be symbolized $\alpha_1(A)$, $\beta_1(A)$, $\gamma_1(A)$, $\delta_1(A)$ respectively. $\alpha_1(A)$ is therefore identically equal to A, $\beta_1(A)$ to 1, $\gamma_1(A)$ to 0, $\delta_1(A)$ to the negative of A. The functions $\alpha_1(A)$ and $\delta_1(A)$ are evidently self-dual, while $\beta_1(A)$ and $\gamma_1(A)$ are duals of each other. $\beta_1(A)$ and $\gamma_1(A)$ admit the expansion 1:0, 0:1 respectively in addition to their complete expansions.

If all of the arguments of a membership function f(B,D,... ..H) be replaced by A, the resulting first order function f(A, A,...A) must be one of the above four first order functions. It will be called the **associated first order function**; and according as it is $\alpha_1(A)$, $\beta_1(A)$, $\gamma_1(A)$, or $\delta_1(A)$, f(B,D,...H) will be said to be an α, β, γ, or δ-function. By referring to §7 we readily verify the following equivalences.

α-function: the uniform terms of the first kind are in the first expression, the uniform terms of the second kind are in the second expression of an expansion of the function;

β-function: the uniform terms of both kinds are in the first expression of an expansion of the function;

γ-function: the uniform terms of both kinds are in the second expression of an expansion of the function;

δ-function: the uniform terms of the first kind are in the second expression, the uniform terms of the second kind are in the first expression of an expansion of the function.

Clearly, the dual of a β-function is a γ-function and conversely, while the dual of an α-function

is an α-function, of a δ-function a δ-function.
It follows in particular that a self-dual function can
only be an α-function or a δ-function.

Unless otherwise stated, by function we shall henceforth
mean a membership function of a finite number of variables in
V', and by closed system a system of membership functions iteratively closed with respect to V'. The above definitions serve
to divide the complete system of functions into four mutually
exclusive classes of functions. In terms of this classification
the iterative properties of uniform terms can be expressed as
follows. The system of all α and β-functions is closed, the
system of all α and γ-functions is closed, the system of all
α-functions is closed. It turns out in the sequel that, except
for the complete system, these three closed systems are the only
ones which contain one or more of the above classes of functions
in their entirety.

Since there are exactly four first order functions of any
one variable, the system of all first order functions is consti-
tuted as follows.

$$[\alpha_1(A), \beta_1(A), \gamma_1(A), \delta_1(A), \alpha_1(B), \beta_1(B), \gamma_1(B), \delta_1(B), \alpha_1(C), \beta_1(C),$$
$$\gamma_1(C), \delta_1(C), \ldots \ldots].$$

This complete system of first order functions is evidently
closed. For every iterative process involving only first order
functions results in a first order function. Its associated contracted system is the finite set of functions $[\alpha_1(A), \beta_1(A), \gamma_1(A),$
$\delta_1(A)]$. Now it is readily seen that the contracted systems associated with closed systems of first order functions are those
subsets of this set which are themselves iteratively closed with
respect to A. Furthermore, it can be directly verified that an
iterative process $f(g(A))$ yields either the function $f(A)$, or
$g(A)$, with the following exceptions.

$$\delta_1(\beta_1(A)) = \gamma_1(A), \quad \delta_1(\gamma_1(A)) = \beta_1(A), \quad \delta_1(\delta_1(A)) = \alpha_1(A).$$

It easily follows that of the fifteen existing subsets of the set of functions $[\alpha_1(A), \beta_1(A), \gamma_1(A), \delta_1(A)]$ exactly nine are contracted closed systems. The corresponding closed systems of first order functions can be symbolized and described as follows:

$$O_1:[\alpha_1], \quad O_2:[\beta_1], \quad O_3:[\gamma_1], \quad O_4:[\alpha_1,\delta_1], \quad O_5:[\alpha_1,\beta_1], \quad O_6:[\alpha_1,$$
$$\gamma_1], \quad O_7:[\beta_1,\gamma_1], \quad O_8:[\alpha_1,\beta_1,\gamma_1], \quad O_9:[\alpha_1,\beta_1,\gamma_1,\delta_1],$$

where $O_4:[\alpha_1,\delta_1]$, for example, means that O_4 consists of all functions $\alpha_1(X)$, $\delta_1(X)$ with $X = A,B,C,\ldots$. As we do not consider the null system of functions a closed system, O_1 to O_9 are the only closed systems of first order functions. Clearly, O_1, O_4, O_7, O_8, and O_9 are self-dual systems, while O_2 and O_3 and O_5 and O_6, are duals of each other. Since the contracted closed system associated with any of these systems itself constitutes a finite set of generators of these systems, these systems are all of order one.

Given any closed system, every iterative process which involves only first order functions that belong to the system must yield a first order function that also belongs to the system. It follows that the first order functions possessed by a closed system themselves constitute a closed system. This closed system, which must be one of the above nine first order systems, will be called the associated first order system. Evidently, the first order function associated with any function in a closed system must itself be in the system, and hence in the associated first order system.

The associated first order system therefore not only specifies the first order functions that belong to the given closed system but also indicates which of the four classes of functions α, β, γ, δ are represented in the closed system.

Referring to the constitution of the nine first order systems, we see that every closed system can be described as an $[\alpha],[\beta],[\gamma],[\alpha,\delta],[\alpha,\beta],[\alpha,\gamma],[\beta,\gamma],[\alpha,\beta,\gamma]$ or $[\alpha,\beta,\gamma,\delta]$ system. Thus, to say that a closed system is an $[\alpha,\beta]$ system means that its associated first order system is $O_5:[\alpha_1,\beta_1]$,

and hence that it consists wholly of α and β-functions and
actually possesses at least the first order α and β-functions.

It is interesting to note that for each closed
first order system there is a maximal closed system with
which it is associated, i.e., one containing all closed
systems with the given associated first order system.[39]

Thus, the set of all functions that belong to $[\alpha,\delta]$ systems may be considered to be an infinite set of generators which
generate some closed system F. If f(B,D,...H) is any one of
these generators, and M,N,...W represent A, $\alpha_1(A)$, or $\delta_1(A)$
then f(M,N,...W) must represent either $\alpha_1(A)$ or $\delta_1(A)$. For
f(B,D,...H) is in some closed $[\alpha,\delta]$ system. As $\alpha_1(A)$ and
$\delta_1(A)$ must also be in this system, the same must be true of the
function of A, f(M,N,...W). But the only functions of A in
a closed $[\alpha,\delta]$ system are $\alpha_1(A)$ and $\delta_1(A)$. It follows by
induction that the infinite set of generators can generate no
other functions of A than $\alpha_1(A)$ and $\delta_1(A)$. F is therefore
an $[\alpha,\delta]$ system; and by its definition it contains all $[\alpha,\delta]$
systems. The above found closed systems consisting of all α,
α and β, α and γ-functions may be called the complete $[\alpha]$,
$[\alpha,\beta]$, $[\alpha,\gamma]$ systems respectively. The complete $[\alpha]$ system
must contain all closed systems consisting wholly of α-functions
and hence is the maximal $[\alpha]$ system. Likewise the complete
$[\alpha,\beta]$ and $[\alpha,\gamma]$ systems are the maximal $[\alpha,\beta]$ and $[\alpha,\gamma]$
systems respectively. The maximal $[\alpha,\beta,\gamma,\delta]$ system is of course
the complete system itself. The composition of the remaining five

---

[39]   This theorem and subsequent proof applies to any iteratively
closed system within the generality of §1 provided the condition
that the number of generators be finite is dropped from the definition of a closed system of given finite order. More generally
and with the same widening of definition, the n-th and lower
order functions of any iteratively closed system (in the sense of
§1) will be the same as the n-th and lower order functions of a
unique closed system of finite order less than or equal to n;
and of all closed systems thus associated for given n with a
closed system of finite order, there will be a maximal closed
system, i.e., one containing all. Though this maximal closed
system theorem is of later origin, the relationship from which
it stems was basic in our original solution of the main problem
of the present paper.

maximal systems must be left to later developments.

This classification of all closed systems according to their associated first order systems is the backbone of the ensuing discussion. This discussion is considerably shortened by the use of the concept of duality. Note that the dual of an [α], [α, δ], [β,γ], [α,β,γ], and [α,β,γ,δ] system is a system of the same type, while the dual of a [β] system is a [γ] system, and conversely, the dual of an [α,β] system an [α,γ] system, and conversely. The derivation of all closed [α], [α,β], and [α,γ] systems is particularly involved. There is therefore considerable value in the fact that after certain non-self-dual [α] and all [α,β] systems have been found, the duals of these systems immediately give us the remaining non-self-dual [α], and all [α,γ] systems.

§11

SYSTEMS OF FUNCTIONS REDUCIBLE TO FIRST ORDER

If a function admits the same expansion as a first order function, it will be said to be reducible to that first order function, or, more briefly, reducible to first order. If E is the argument of the first order function, then according as the given function is an α, β, γ, or δ-function, the first order function, and hence the given function, admits the expansion E:e, 1:0, 0:1, e:E respectively. In the first and last case E is the only relevant variable of the given function, while in the second and third case all of the arguments of the given function are irrelevant. In the latter two cases the argument of the first order function can be chosen arbitrarily.

The condition that a function be reducible to first order is iterative. For if f(M,N,...W) is an iterative process depending on variables A,B,...I, and built up out of functions reducible to first order, then f(M,N,.. ...W) admits some expansion R:r, 1:0, 0:1, r:R in terms of M,N,...W, and hence some expansion E:e, 1:0, 0:1, e:E in terms of A,B,...I. The system of all functions reducible to first order is therefore closed.

This complete system of functions reducible to first
order is composed as follows.

For each set of n variables it possesses n α-functions
of those variables, one β-function, one γ-function, and n δ-
functions. This follows from the fact that while all the vari-
ables are irrelevant in the case of a β and γ-function, any
one of them can be chosen as the relevant·variable of an α and
δ-function. The associated contracted system therefore possesses exactly $2n+2$ functions of order n for every positive
integer n.

Consider any closed system of functions reducible to first
order which does not consist wholly of first order functions.
Also consider an arbitrary finite set of variables. By §1 the
system will have at least one function M on this set of variables. If the system has a β-function, it will also have the
function $\beta_1(M)$. Hence, if the system has one β-function, it
must have all β-functions reducible to first order. Likewise
for γ-functions. If the system has an α-function of order
greater than one, by replacing the irrelevant variables of this
function by M, its relevant variable by any variable E in
the given set of variables, an α-function on the given set of
variables is obtained with relevant variable E. Hence, if the
system has one α-function of order greater than one, it must
have all α-functions reducible to first order. Since $\delta_1(M)$
is an α-function when M is a δ-function, and conversely, it
also follows that if the system has a δ-function, and at least
one α or δ-function of order greater than one, then it has all
α and δ-functions reducible to first order. On the other hand,
the system may have no α or δ-functions other than first order
functions. For, if f is any β or γ-function reducible to
first order, f(M,N,...W) is also a β or γ-function. While
if f is a first order α or δ-function, f(M) will be a β
or γ-function if M is a β or γ-function, a first order α
or δ-function if M is a variable, or a first order α or δ-
function.

It follows that for each associated first order system there
are at most two closed systems of functions reducible to first
order, yet not of the first order, one system possessing all

functions reducible to first order of the types indicated by the
associated first order system, the other possessing all such β
and γ-functions, but only first order α and δ-functions. In
describing these systems we shall use the symbols $\overset{*}{\alpha}_1$, $\overset{*}{\beta}_1$, $\overset{*}{\gamma}_1$,
$\overset{*}{\delta}_1$, to mean any α, β, γ, δ-function respectively that is
reducible to first order. If a closed system consists of functions reducible to first order, we shall say that the closed
system is reducible to first order. By scanning the nine possible associated first order systems, we find that there are exactly thirteen closed systems reducible to first order, yet not
of the first order. They may be symbolized and described as follows,

$$R_1:[\overset{*}{\alpha}_1], \quad R_2:[\overset{*}{\beta}_1], \quad R_3:[\overset{*}{\gamma}_1], \quad R_4:[\overset{*}{\alpha}_1,\overset{*}{\delta}_1], \quad R_5:[\alpha_1,\overset{*}{\beta}_1], \quad R_6:[\overset{*}{\alpha}_1,$$

$$\overset{*}{\beta}_1], \quad R_7:[\alpha_1,\overset{*}{\gamma}_1], \quad R_8:[\overset{*}{\alpha}_1,\overset{*}{\gamma}_1], \quad R_9:[\overset{*}{\beta}_1,\overset{*}{\gamma}_1], \quad R_{10}:[\overset{*}{\alpha}_1,\overset{*}{\beta}_1,\overset{*}{\gamma}_1],$$

$$R_{11}:[\overset{*}{\alpha}_1,\overset{*}{\beta}_1,\overset{*}{\gamma}_1], \quad R_{12}:[\alpha_1,\overset{*}{\beta}_1,\overset{*}{\gamma}_1,\delta_1], \quad R_{13}:[\overset{*}{\alpha}_1,\overset{*}{\beta}_1,\overset{*}{\gamma}_1,\overset{*}{\delta}_1].$$

Clearly, α and δ-functions reducible to first order are
self-dual, while the dual of a β-function reducible to first order is a γ-function reducible to first order, and conversely.
Hence R_2 and R_3, R_5 and R_7, R_6 and R_8 are dual closed
systems, the others self-dual. Since in the above argument the
function M can be generated by any second order function in
the system, while the α or δ-function of order greater than
one can be any second order α or δ-function in the system, it
follows that each of the above systems can be generated by first
and second order functions, and hence is itself of the second
order. Together with the first order systems there are therefore 22 closed systems reducible to first order.

§12

[β], [γ], AND [β,γ] SYSTEMS

Suppose that a function f(A,B,...I) in a closed [β] system has a term t(A,B,...I) in the second expression of its complete expansion. Since f(A,B,...I) can only be a β-function,

the uniform terms AB...I and ab...i must both be in the first
expression of this complete expansion. t(A,B,...I) therefore
must have at least one capital letter, and at least one small
letter. Now in the function f(A,B,...I) replace each argument
E by β_1(A) or a according as E is capital or small in
t(A,B,...I). That is, in the complete expansion of f(A,B,...I)
replace E:e by 1:0 or A:a in the manner indicated. t(A,B,
...I) then reduces to a, AB...I to A, and hence the complete expansion of f(A,B,...I) to A:a. That is, the above
iterative process, which involves only functions in the given
closed [β] system, would yield the function α_1(A). α_1(A)
would therefore belong to the given closed [β] system, which
is impossible.

Hence, all functions in a closed [β] system have 0 for
the second expressions of their complete expansions, and therefore admit the expansion 1:0. Dually, all functions in a closed
[γ] system admit the expansion 0:1. By the same argument the
functions in a closed [β,γ] system admit expansions 1:0 and
0:1.

That is, every closed [β], [γ], and [β,γ] system
consists wholly of functions reducible to first order.
By referring to the last two sections, we thus see that
the only closed [β] systems are O_2 and R_2, the
only closed [γ] systems are O_3 and R_3, the only
closed [β,γ] systems are O_7 and R_9. It follows
incidentally that the maximal [β], [γ] and [β,γ]
systems are R_2, R_3, and R_9 respectively.

This result, though trivial in itself, is significant as
marking the first step in our general plan of determining all
closed systems through their associated first order systems.

§13

LOGICAL SUM SYSTEMS; LOGICAL PRODUCT SYSTEMS

The logical sum of two or more distinct variables A,B,...I
is a function having the simple expansion A+B+...+I:ab...i. If

a function admits the same expansion as such a 'logical sum function' it will be said to be reducible to that logical sum function. The arguments of the logical sum function are relevant variables of the given function, all other arguments of the given function, irrelevant. Obviously, all functions reducible to logical sum functions are α-functions.

Now it is easily verified that the system consisting of all functions reducible to first order α, β, and γ-functions, and to logical sum functions is closed. We proceed to determine the closed subsystems of this complete 'logical sum system'.

If $f(A,B)$ is the function with expansion A+B:ab, then $f(A,A)$ has the expansion A:a, $f(A,f(B,...f(H,I)...))$ the expansion A+B+...+I:ab...i. Furthermore, if M and N are variables or first order α-functions or logical sum functions, $f(M,N)$ is also a first order α-function or logical sum function. Hence, $f(A,B)$ generates S_1, the closed system consisting of all first order α-functions and all logical sum functions.

Now any closed system of the type we are considering which does not consist wholly of functions reducible to first order must possess a function reducible to a logical sum function. If one of the relevant variables of this function be replaced by A, all of the remaining variables by B, the above function $f(A,B)$ with expansion A+B:ab results. The closed system in question therefore contains S_1.

If, furthermore, this closed system possesses an α-function not in S_1, this α-function will have at least one relevant, and at least one irrelevant variable. If the relevant variables of this function are replaced by an arbitrary function M in S_1, the irrevelant variables by an arbitrary function N in S_1, the resulting function, which will also be in the closed system, is a function reducible to M. It follows that the closed system must then possess all functions reducible to functions in S_1, i.e., all the α-functions possessed by the complete logical sum system.

There remains to see the effect of the presence of β and γ-functions. As in the case of the R systems, the closed system under discussion must possess all functions reducible to first order β and γ-functions therein. If the system does have a γ-function, by replacing B by $\gamma_1(B)$, i.e., B:b by

0:1. $f(A,B)$ with expansion A+B:ab becomes a function $g(A,B)$ with expansion A:a. The system therefore has the α-function $g(A,B)$ with relevant variable A, irrelevant variable B, and hence possesses all α-functions reducible to functions in S_1. On the other hand, the presence of β-functions in the system is consistent with all of its α-functions being in S_1. Thus, if any variable in a logical sum function be replaced by a β-function with expansion 1:0, the expansion of the logical sum function also reduces to 1:0, and but a β-function results.

The various possibilities are now easily seen. We shall continue the symbolism used for the R systems, and in addition use σ to symbolize an arbitrary logical sum function, $\overset{*}{\sigma}$ an arbitrary function reducible to a logical sum function. Then the only closed subsystems of the complete logical sum system other than closed systems reducible to first order are the following.[40]

$$S_1:[\alpha_1,\sigma], \quad S_2:[\overset{*}{\alpha}_1,\overset{*}{\sigma}], \quad S_3:[\alpha_1,\overset{*}{\beta}_1,\sigma], \quad S_4:[\overset{*}{\alpha}_1,\overset{*}{\beta}_1,\overset{*}{\sigma}], \quad S_5:[\overset{*}{\alpha}_1,\overset{*}{\gamma}_1,$$

$$\overset{*}{\sigma}], \quad S_6:[\overset{*}{\alpha}_1,\overset{*}{\beta}_1,\overset{*}{\gamma}_1,\overset{*}{\sigma}].$$

The above derivation shows that each of these 'logical sum systems', can be generated by a suitable choice of generators from the functions $\beta(A)$, $\gamma_1(A)$, $f(A,B)$ with expansion A+B:ab, $g(A,B)$ with expansion A:a. <u>Hence they are all of the second order</u>.

The dual of the expansion A+B+...+I:ab...i is the expansion AB...I:a+b+...+i. Hence the dual of a logical sum function is a 'logical product function', i.e., a function which is the logical product of two or more distinct variables, while the dual of a function reducible to a logical sum function is a function 'reducible to a logical product function'. By the principle of duality we may then immediately write down the closed systems which have no other functions than those reducible to first order α, β, and γ functions, and to logical product functions, and which are not themselves reducible to first order. Introducing the symbols π and $\overset{*}{\pi}$ as duals of σ and $\overset{*}{\sigma}$, we thus have the following six 'logical product

---

[40]   Allowing a closed system to be an improper subsystem of itself.

systems' which, in order, are the duals of the above six logical
sum systems, and <u>which are also of the second order</u>.

$$P_1 : [\overset{}{\alpha}_1, \pi \,], \quad P_2 : [\overset{*}{\alpha}_1, \overset{*}{\pi}\,], \quad P_3 : [\alpha_1, \overset{*}{\gamma}_1, \pi\,], \quad P_4 : [\overset{*}{\alpha}_1, \overset{*}{\gamma}_1, \overset{*}{\pi}\,], \quad P_5 :$$

$$[\overset{*}{\alpha}_1, \overset{*}{\beta}_1, \overset{*}{\pi}\,], \quad P_6 : [\overset{*}{\alpha}_1, \overset{*}{\gamma}_1, \overset{*}{\beta}_1, \overset{*}{\pi}\,].$$

§14

[α,β,γ] SYSTEMS

By referring to §8 we see that the only functions satisfying the [A:a] condition are β and γ-functions reducible to
first order, and α-functions which admit a normal [A:a] expansion. Since the [A:a] condition is iterative, the system
of all [A:a] functions is closed. We thus have the closed
[α,β,γ] system symbolized and described as follows.

A_1. All functions satisfying the [A:a] condition.

Now suppose that a function belonging to a closed [α,β,γ]
system does not satisfy the [A:a] condition. Then, as seen in
§8, there will be a term in the first expression and a term in
the second expression of the complete expansion of this function
with respect to which the arguments of the function fall into
at most three groups as follows:

   term in the first expression : capital small small,
   term in the second expression : capital small capital.

Clearly, the third group cannot be empty, or the two terms would
not be distinct. Now replace the variables in the first group
by $\beta_1(A)$, in the second by $\gamma_1(A)$, in the third by A. The
two terms become a and A respectively, hence the complete
expansion a:A, and the function $\delta_1(A)$. The given closed system would therefore possess the function $\delta_1(A)$, which contra-
dicts its being an [α,β,γ] system.

It follows that every closed [α,β,γ] system satisfies

the [A:a] condition, i.e., consists wholly of functions satisfying the [A:a] condition.

Hence, also, every closed [α,β,γ] system is contained in A_1 which is therefore the maximal [α,β,γ]
system.

Since the first expression of the normal expansion of an
[A:a]α-function is the logical sum of logical products of capital letters, it follows that every [A:a]α-function can be generated by the two [A:a] functions A+B and AB. Hence A_1
can be generated by $β_1(A)$, $γ_1(A)$, A+B, and AB.

A_1 is therefore of the second order. It further
follows that A_1 is the only closed [α,β,γ] system
having both of the functions A+B and AB.

Now let a closed [α,β,γ] system possess an α-function
whose normal [A:a] expansion has in its first expression a term
CE...F with more than one letter. Since CE...F would not be
in the normal expansion if it were contained in another term of
the expansion, every other term of this first expression involves
some capital letter other than C,E,...F. By replacing every argument of the given function other than C,E,...F by $γ_1(A)$,
i.e., the corresponding capital letters by 0, the first expression in question reduces to CE...F. By further replacing C
by A, and E,...F by B, the first expression becomes AB.
That is, the closed system possesses the function with expansion
AB:a+b, i.e., the function AB. By the dual argument we see
that if a closed [α,β,γ] system possesses an α-function whose
normal expansion has in its second expression a term with more
than one letter, it also possesses the function with expansion
A+B:ab, i.e., the function A+B.

If then a closed [α,β,γ] system is not to be A_1, and so
is not to possess both AB and A+B, either the normal expansion of its α-functions all have first expressions with terms
of one letter only, or second expressions with terms of one letter only. That is, either its α-functions are all reducible to
first order functions and logical sum functions, or they are all
reducible to first order functions and logical product functions.

As the β and γ-functions in the system must be reducible to
first order, it follows that the system can then only be an O,
R, S, or P system. We can therefore conclude that the only
$[\alpha,\beta,\gamma]$ systems are O_8 , R_{10} , R_{11} , S_6 , P_6 , and A_1 .

§15

ALTERNATING SYSTEMS

A function will be said to satisfy the <u>alternating condi-
tion</u> if replacing an argument of the function by its negative
either leaves the function unchanged, or changes it into its
negative.[41] By §4 the arguments of the first kind are the irrelevant variables of the function, those of the second kind
its relevant variables. The set of relevant variables of an
alternating function will be called its <u>relevant subset</u>. Clearly, if any number of arguments of an alternating function are
changed into their negatives the function is left unchanged, or
is turned into its negative, according as an even, or odd, number of variables in its relevant subset have been changed into
their negatives. It follows that <u>the alternating condition is
iterative</u>. For if $f(M,N,...W)$ is an iterative process built
up out of alternating functions and depending on variables A,
B,...I, changing any one of these variables into its negative
leaves M,N,...W unchanged or turned into its negative, and
hence leaves $f(M,N,...W)$ unchanged or turned into its negative.

From the definition of an alternating function we see that
a complete expansion corresponds to an alternating function when
and only when interchanging any capital and corresponding small
letter throughout the expansion either leaves the expansion unchanged, or has the effect of interchanging the two expressions
of the expansion. Since in the first case the letter in ques-

---

[41]   For truth-functions a typical example of an 'alternating
function' is $p \equiv q$. Thus, reinterpreted as closed systems of
truth-functions, L_2 below may be said to be generated by
equivalence, L_3 by the dual, which is also the negative, of
equivalence, L_1 by, for example, equivalence and negation.

tion corresponds to an irrelevant variable of the function, in
the second to a relevant variable, it follows that two terms of
the complete expansion of an alternating function which differ
in the capitalization of but a single letter are in the same or
different expressions of the expansion according as the letter
corresponds to an irrelevant or relevant variable of the function. By successive application of this special result we obtain the following general result. Two terms of the complete
expansion of an alternating function which differ in the capitalization of any number of letters are in the same or different expressions of the expansion according as an even or odd
number of those letters correspond to variables in the relevant
subset of the function.

        In particular, a term of the complete expansion of
an alternating function $f(A,B,...I)$ is in the same
expression as $AB...I$, or in the opposite expression,
according as an even or odd number of the small letters
in the term correspond to variables in the relevant subset of $f(A,B,...I)$.

An alternating function $f(A,B,...I)$ is therefore determined by its relevant subset, and the expression of its complete
expansion in which $AB...I$ appears. Corresponding to any set
of variables $A,B,...I$, and subset of it chosen as the relevant
subset, there are thus but two alternating functions, one with
$AB...I$ in the first expression of its complete expansion, the
other with $AB...I$, in the second expression.

    Dual terms differ in the capitalization of all their let-
ters. Hence, when the relevant subset of an alternating function has an even number of members, dual terms of the complete
expansion of the function are always in the same expression,
when odd, in different expressions.

        Alternating functions with odd relevant subsets
are therefore self-dual, and conversely. By considering
the dual terms $AB...I$ and $ab...i$ we further see
that alternating functions with even relevant subsets
are β or γ-functions, those with odd relevant sub-

sets α or δ-functions.

It is interesting to note that if an expansion of an alternating function has an incomplete term, the letters missing in
that term must correspond to irrelevant variables of the function. For among the terms of the complete expansion of the
function which are contained in the incomplete term, and which
are therefore in the same expression of the complete expansion,
there will be at least one pair which differ only in the capitalization of any chosen one of those letters. Now we saw in
§4 that a function admits an expansion involving only its relevant variables. In the case of an alternating function, therefore, each term of such an expansion must be a complete term on
the relevant variables. If, then, we exclude the trivial case
of expansions in which the same term appears more than once, we
can conclude that an alternating function admits but one expansion involving only its relevant variables, and that is a complete expansion on those variables.

The same result of §4 shows that any function, and hence
any alternating function, with not more than one relevant variable is reducible to first order. Conversely, it can be directly
verified that any function reducible to first order is an alternating function with not more than one relevant variable.

That is, the class of functions reducible to first
order is identical with the class of alternating functions having not more than one relevant variable. By
examining the expansion of an alternating function which
involves only the relevant variables of the function,
we can then prove the following fact. An alternating
function which satisfies any of the three conditions
[A:a], [AA:], [:aa] must be reducible to first order.

We turn now to the closed systems of alternating functions.
As closed systems of functions reducible to first order have al-
ready been studied, it will be assumed that the systems under
discussion have a function with relevant subset of more than one
variable. Our derivation will fall into two parts according as
the system has, or has not, a function with even relevant subset.

We shall constantly, though tacitly, be using the previously
proved fact that an iterative process built up out of alternating functions yields an alternating function. The relevant subset of this function can in each instance be found by a direct
application of the definition of an alternating function.

Consider then first a closed system of alternating functions
which possesses a function with even relevant subset. By replacing all of the arguments of this function by A, a function g(A)
is obtained with null relevant subset. Now take a function in
the system which has a relevant subset of more than one variable,
and replace two of the variables in this subset by A and B
respectively, all of the remaining arguments of the function by
g(A). We thus obtain a function f(A,B), also in the system,
with relevant subset (A,B).

f(A,A) is a function of A with null relevant subset, f(A,
f(A,A)) with relevant subset (A). If M is any alternating
function of A,B,...H, then f(M,f(I,I)) has the same relevant
subset as M, f(M,I) that of M with I added. By induction
it follows that through f(A,B) one of the two possible alternating functions corresponding to any set of variables A,B,...I,
and any subset of it chosen as the relevant subset, can be constructed.

If f(A,B) has AB in the first expression of its complete
expansion, then, by the iterative properties of uniform terms,
this constructed function is the one with AB...I in the first
expression of its complete expansion. Hence, in this case, the
system possesses all alternating functions of A,B,...I which
have AB...I in the first expressions of their complete expansions. That is, the system possesses all α and β-alternating
functions. Now γ and δ-alternating functions are the negatives of β and α-alternating functions respectively. Hence,
if the system also possesses a δ-function, it will possess all
γ and δ-alternating functions in addition to all α and β-
alternating functions. The same is true if it possesses a γ-
function. For, if we replace B by $\gamma_1(A)$, the complete expansion AB+ab:Ab+aB of f(A,B) becomes the complete expansion
a:A of $\delta_1(A)$. Hence, if the system does not consist of the
α and β-alternating functions, it possesses all alternating
functions. When f(A,B) has AB in the second expression of

its complete expansion, it has ab also in the second expression,
so that the dual results follow. We can therefore conclude that
the following are the only closed systems of alternating functions which do not consist wholly of functions reducible to first
order, and which possess a function with even relevant subset.

L_1. All alternating functions.

L_2. All α and β alternating functions.

L_3. All α and γ alternating functions.

That these systems are closed follows from the iterative character of the alternating condition and the iterative properties of
uniform terms.

Consider now a closed system of alternating functions which
has only functions with odd relevant subsets. In a function in
the system with odd relevant subset of more than one variable replace two of the relevant variables by A and B respectively,
all the remaining arguments of the function by C. We thus obtain a function f(A,B,C), also in the system, with relevant
subset (A,B,C).

f(E,E,E) is a function of E with relevant subset (E).
If M is any alternating function with odd relevant subset, then
if G is not an argument of M, f(M,G,G) has the same relevant
subset as M, while if H and J are not arguments of M,
f(M,H,J) has for relevant subset that of M with H and J
added. It follows by induction that through f(A,B,C) one of
the two possible alternating functions corresponding to any set
of variables, and any odd subset of it chosen as relevant, can
be constructed.

We have seen that the alternating functions with odd relevant subsets are the α and δ alternating functions. What-
ever function is constructed through f(A,B,C) in the above
manner, the negative of that function is the other possible alternating function corresponding to the given set of variables,
and odd relevant subset thereof. Hence, if the system has a δ-
function, it will have all alternating functions with odd relevant subsets, i.e., all α and δ alternating functions.

Otherwise the system can only have α-functions. $f(A,B,C)$ is then an α-function so that each of the above functions constructed by means of $f(A,B,C)$ is an α-function. But only one of the two possible alternating functions corresponding to a given set of variables and odd relevant subset thereof can be an α-function. The system therefore has all of the alternating functions with odd relevant subsets that are α-functions, i.e., it has all α alternating functions. Hence the only possible closed systems of alternating functions which do not consist wholly of functions reducible to first order, and which do not have a function with even relevant subset, are the following.

L_4. All α alternating functions.

L_5. All α and δ alternating functions.

That L_4 is closed is seen immediately. On the other hand, L_5 can be described as the system of all self-dual alternating functions, and hence is closed because of self-duality, and the alternating condition, are iterative.

Our derivation shows that L_1, L_2, and L_3 can be generated by the function $f(A,B)$ with the possible addition of $\delta_1(A)$.

Hence L_1, L_2, and L_3 are of the second order.

Since a second order alternating function with odd relevant subset is reducible to first order, L_4 and L_5 cannot be of the second order. They can however be generated by the above mentioned $f(A,B,C)$ with the possible addition of $\delta_1(A)$.

Hence L_4 and L_5 are of the third order.

Note that L_2 and L_3 are dual systems, while L_1, L_4 and L_5 are self-dual, the last two in fact consisting wholly of self-dual functions.

L_1, L_2, L_3, L_4, L_5 are then the only closed systems of alternating functions which do not consist wholly of functions

reducible to first order. The O, R, and L systems therefore constitute all of the closed alternating systems, i.e.,
closed systems of alternating functions.

§16

FAILURE OF THE ALTERNATING CONDITION

An argument E of an alternating function f(A,B,...I) is
evidently an irrelevant or relevant variable of the function according as the term whose only small letter is e is in the same
expression of the complete expansion of f(A,B,...I) as AB...I
or in the opposite expression. Hence, two terms of the complete
expansion of an alternating function f(A,B,...I) which differ
in the capitalization of but a single letter E are in the same
or different expressions of the expansion according as AB...I
and the term whose only small letter is e are in the same or
different expressions. Conversely, a complete expansion which
satisfies this condition for each of the letters on which it is
written must correspond to an alternating function. For, on interchanging any capital and corresponding small letter in the expansion, either every term of the expansion is changed into a
term of the same expression, or every term is changed into a term
of the opposite expression, i.e., either the expressions are unchanged or interchanged.

In the complete expansion of a function f(A,B,...I) let
t_0 and t_1 respectively be the uniform term AB...I and a
term with but one small letter, t_2 and t_3 any other pair of
terms which differ only in the capitalization of that letter.
Then, if f(A,B,...I) does not satisfy the alternating condition, for some quadruplet (t_0, t_1, t_2, t_3) either t_0 and t_1
are in the same expression, t_2 and t_3 in different expressions, or the reverse. That is, for some quadruplet $(t_0, t_1, t_2,
t_3)$ three terms are in one expression, the fourth in the other.
Now the variables A,B,...I fall into at most three groups with
respect to their capitalization in those terms, to wit,

t_0			capital	capital	capital
t_1			capital	capital	small
t_2	or	t_3	capital	small	capital
t_3	or	t_2	capital	small	small

While no variable may belong to the first group, the last two
groups cannot be empty or the four terms would not be distinct.
Now in the function $f(A,B,...I)$ replace any of the arguments
that may fall in the first group by A, those in the second
and third groups by B and C respectively. We thus either
get a function $g(B,C)$ with three terms of its complete expansion in one expression, the fourth in the other, or a function
$h(A,B,C)$ with an A-component of this type. Hence the following.

   If a closed system possesses a function which
fails to satisfy the alternating condition, then it
also has either a second order function with an odd number of terms in each expression of its complete expansion, or a third order function whose A-component is of
this type.
   If, furthermore, the system possesses a β-function
we can modify the above argument by replacing the variables that may fall in the first group by $β_1(B)$, and
thus be sure to get the second order function in question.

§17

[α, β , γ , δ] SYSTEMS

   The maximal closed [α, β, γ, δ] system is the
complete system

   C_1. All functions.

Since the first expression of an expansion of a function
is the logical sum of logical products of variables and their
negatives, we see that the functions with expansions a:A, AB:
a+b, A+B:ab constitute a set of generators of C_1.

It follows that C_1 is a closed system of the second order.

We now prove that $\delta_1(A)$ and any function $f(A,B)$ whose complete expansion has a one term expression generate C_1.

Note that $\delta_1[f(A,B)]$ has a complete expansion whose other expression consists of a single term. Now in the function with the one term first expression replace each argument that appears as a small letter in that term by its negative; in the function with the one term second expression replace each argument that appears as a capital letter in that term by its negative. The one term first expression thus becomes AB, the one term second expression ab. That is, by means of $\delta_1(A)$ and $f(A,B)$ we can construct the functions with expansions $AB:a+b$ and $A+B: ab$. As these functions together with $\delta_1(A)$ generate C_1, $\delta_1(A)$ and $f(A,B)$ also generate C_1.

Since an $[\alpha, \beta, \gamma, \delta]$ system must have a β-function, it follows from the preceding section that every closed $[\alpha, \beta, \gamma, \delta]$ system which does not consist wholly of alternating functions possesses such a second order function $f(A,B)$. Since it also possesses a δ-function, it must contain C_1, and hence is C. Every closed $[\alpha, \beta, \gamma, \delta]$ system other than C_1 is therefore an alternating system. It follows that the only closed $[\alpha, \beta, \gamma, \delta]$ systems are O_9, R_{12}, R_{13}, L_1, and C_1.

Pausing a moment to review our progress in determining all closed systems through their associated first order systems, we see that there remains the complete determination of all closed $[\alpha]$, $[\alpha, \beta]$, $[\alpha, \gamma]$, and $[\alpha, \delta]$ systems. The corresponding discussion constitutes the most significant part of our development.

§18

[α] SYSTEMS; PRELIMINARY DISCUSSION

An α-function $f(A,B,\ldots I)$ has the dual terms $AB\ldots I$ and $ab\ldots i$ in the first and second expressions respectively of its complete expansion. If $f(A,B,\ldots I)$ is not self-dual, it

will then have some other pair of dual terms in one and the same
expression of its complete expansion. Now in the function f(A,
B,...I) replace the arguments that appear as capital letters in
one of these dual terms, and hence as small letters in the other,
by A, those that appear as small letters in the first of these
terms, and hence as capital letters in the second, by B. The
two terms become Ab and aB respectively, and hence the complete expansion AB+Ab+aB:ab or AB:Ab+aB+ab. The function
f(A,B,...I), therefore, either reduces to the function with the
expansion A+B:ab or to the function with the expansion AB:a+b.
It becomes convenient to use an expansion of a function which
involves all of the arguments of the function as a symbol for
the function.

> We can then state that every closed [α] system
> which does not consist wholly of self-dual functions
> possesses either the function A+B:ab or the function
> AB:a+b.

Note that A+B:ab satisfies the [:aa] condition but not
the [AA:] condition, while AB:a+b satisfies the [AA:] condition but not the [:aa] condition. Now suppose that a closed
[α] system which possesses the function A+B:ab fails to satisfy the [:aa] condition, i.e., possesses a function which fails
to satisfy the [:aa] condition. Then some function in the system has a pair of terms in the second expression of its complete
expansion with respect to which the arguments of the function
fall into at most three groups as follows:

> capital capital small
> capital small   capital

The last two groups cannot be empty, or the term consisting of
capital letters only would be in the second expression of this
complete expansion, in contradiction to the function in question
being an α-function. Now replace the arguments of this function
that are in the second and third group by A and B respectively, those, if any, that are in the first group by A+B. When
the first group is not empty, the above two terms become (A+B)Ab

and (A+B)aB, and hence, in any case, reduce to Ab and aB. The function therefore reduces to the function AB:Ab+aB+ab, i.e., to the function AB:a+b. The given closed [α] system therefore possesses the function AB:a+b. By the dual argument, if a closed [α] system possesses the function AB:a+b, but does not satisfy the [AA:] condition, then it also must have the function A+B:ab.

Hence, every closed [α] system which does not consist wholly of self-dual functions and which does not satisfy either the [:aa] or the [AA:] condition possesses both the function A+B:ab and the function AB:a+b.

By referring to §14 we see that the functions A+B:ab, AB:a+b generate the <u>second order</u> closed system:

A_4. All α-functions that satisfy the [A:a] condition.

We now prove that the only other closed [α] system which possesses both of the functions A+B:ab, AB:a+b is the complete [α] system:

C_4. All α-functions.

The system in question must have an α-function which fails to satisfy the [A:a] condition. There will therefore be a term in the first expression and a term in the second expression of the complete expansion of this function with respect to which the arguments of the function fall into at most three groups as follows:

```
term in the first  expression : capital small small
term in the second expression : capital small capital
```

The third group cannot be empty, or the two terms would not be distinct; the first and second group cannot be empty, or the function would not be an α-function. Now replace the arguments of the function that are in the first group by A+B+...+I, those that are in the second group by AB...I, those that are in the third group by any function M in the given system whose arguments constitute a subset of the set of variables (A,B,...I). The two terms in question become (A+B+...+I)(a+b+...+i)m, (A+B+...+I)(a+b+...+i)M respectively, the given function a function of A,B,...I which may be written g(A,B,...I,M), and which is also in the given system. Now every complete term on A,B,...I other than AB...I and ab...i is contained in (A+B+...+I)(a+b+...+i). Hence g(A,B,...I,M) has the property that every complete term t(A,B,...I) other than AB...I and ab...i is contained in corresponding expressions of expansions of g(A,B,...I,M) and $\delta_1(M)$.

By a result of the preceding section, we see that $\delta_1(A)$ and $\pi(A,B)$, the logical product of A and B, generate the complete system C_1. Consider then any α-function f(A,B,...I), and any mode of building up this α-function by means of $\delta_1(A)$ and $\pi(A,B)$. Such a process is an operation in the sense of §1. We now inductively define a corresponding operation built up by means of functions in the system under discussion, and hence yielding a function in this system. Components of rank zero are to be the same in the new operation as in the old. Assume that components of the new operation have been defined which are to correspond to components of rank less than ρ of the old. Then to a component of rank ρ of the old operation which is of the form $\pi(\mathcal{M},\mathcal{N})$ we make correspond a component $\pi(\mathcal{M}',\mathcal{N}')$ of the new operation, where $\mathcal{M}'$ and $\mathcal{N}'$ correspond to $\mathcal{M}$ and $\mathcal{N}$ respectively; to a component of rank ρ of the form $\delta_1(\mathcal{M})$ we make correspond a component g(A,B,...I,$\mathcal{M}'$), where $\mathcal{M}'$ corresponds to $\mathcal{M}$. We thus ultimately obtain the new operation that is to correspond to the old. Now let M, N, M', N' be the variables or functions yielded by the operations $\mathcal{M}$, $\mathcal{N}$, $\mathcal{M}'$, $\mathcal{N}'$ respectively. Then by §5 any complete term t(A,B,...I) which is contained in corresponding expressions of expansions of M and M', N and N', is contained in corresponding expressions of expansions of $\pi(M,N)$ and $\pi(M',N')$. And, as we have just seen, any complete term t(A,B,...I) other than AB...I or ab...i is contained in corresponding expressions of expansions of $\delta_1(M)$ and g(A,B,...I,M), and hence, if contained in corresponding expressions of expansions of M and M', is contained

in corresponding expressions of expansions of $\delta_1(M)$ and $g(A,$
$B,...I,M')$. It follows by induction that every complete term
$t(A,B,...I)$ other than $AB...I$ and $ab...i$ is contained in
corresponding expressions of expansions of $f(A,B,...I)$ — the
function yielded by the old operation — and the function of
$A,B,...I$ yielded by the new operation. But the same must be
true of the complete terms $AB...I$ and $ab...i$ since both
functions are α-functions. It follows that the complete expansions of the functions are identical, and hence the functions
are identical. The new operation therefore yields the same function $f(A,B,...I)$ as the old.

Since $f(A,B,...I)$ is any α-function of $A,B,...I$, the
closed [α] system under discussion possesses all α-functions,
and hence is none other than C_4.

This proof shows incidentally that C_4 can be generated by
A+B:ab, AB:a+b, and any α-function that fails to satisfy the
[A:a] condition. Our observation that all three groups of variables associated with two terms resulting in this failure must
be existent shows that while this non-[A:a] α-function may be
of the third order, every α-function of order less than three
does satisfy the [A:a] condition. It follows that $\underline{C_4}$ is of
the third order.

As a result of this preliminary discussion we can
state that every closed [α] system other than A_4
and C_4 either consists wholly of self-dual functions,
or else satisfies either the [:aa] or the [AA:]
condition.

§19

[α,β] AND [α,γ] SYSTEMS; PRELIMINARY DISCUSSION

As was seen in §10, the closed [α,γ] systems are the duals
of the closed [α,β] systems. Our explicit development may
therefore be restricted to closed [α,β] systems. Since $\beta_1(A)$,
whose complete expansion is A+a:0, satisfies the [:aa] condition, but is not self-dual and does not satisfy the [AA:] condition, a closed [α,β] system may satisfy the [:aa] condition,

but does not consist wholly of self-dual functions and cannot satisfy the [AA:] condition. There results in consequence a certain analogy between the discussion of closed [α,β] systems, and closed [α] systems possessing the function A+B:ab. The purpose of our preliminary discussion is to derive all of the closed [α,β] systems which do not satisfy the [:aa] condition.

As in the case of closed [α] systems some function in such a closed [α,β] system will fail to satisfy the [:aa] condition, and for this function the arguments will fall into at most three groups

$$
\begin{array}{lll}
\text{capital} & \text{capital} & \text{small} \\
\text{capital} & \text{small} & \text{capital}
\end{array}
$$

with respect to some pair of terms in the second expression of the complete expansion of the function. By replacing the arguments of the function that are in the second and third groups, again necessarily present, by A and B respectively, those, if any, that are in the first group by $\beta_1(A)$, the two terms become Ab and aB. As the resulting function may now be either an α or a β-function, the resulting complete expansion is either AB:Ab+aB+ab or AB+ab:Ab+aB.

Hence, every closed [α,β] system which does not satisfy the [:aa] condition possesses either the function AB:a+b or the function AB+ab:Ab+aB.

Consider first a closed [α,β] system which possesses the function AB:a+b, but fails to satisfy the [A:a] condition. Then for some function in the system there will be a term in the first expression and a term in the second expression of the complete expansion of the function with respect to which the arguments of the function fall into at most three groups as follows:

term in the first expression : capital small small
term in the second expression : capital small capital.

While the last two groups cannot be empty, the first group may

now be empty. By replacing the arguments that may fall in the
first group by $\beta_1(A)$, those in the second by AB...I, those
in the third by any function M in the system whose arguments
constitute a subset of the set of variables (A,B,...I), the
two terms in question become (a+b+...+i)m, (a+b+...+i)M respectively. Since every complete term t(A,B,...I) other than
AB...I is contained in a+b+...+i, the resulting function h(A,
B,...I,M) has the property that every complete term t(A,B,...I)
other than AB...I is contained in corresponding expressions
of expansion of h(A,B,...I,M) and $\delta_1(M)$. As in the derivation of the closed [α] system C_4, we therefore conclude that
every function of A,B,...I whose complete expansion has AB...I
in its first expression can be generated by means of h(A,B,...I,
M) and AB:a+b, and hence is in the system under discussion.

Hence, the only closed [α,β] system which possesses the function AB:a+b, but fails to satisfy
the [A:a] condition, is the complete [α,β] system
C_2. All α and β-functions.

As the above function which fails to satisfy the [A:a]
condition may now be a β-function of order two, it follows that
C_2 is of order two.

All other closed [α,β] systems that possess the function
AB:a+b therefore satisfy the [A:a] condition. If the α-functions in such a system all have normal [A:a] expansions with
second expressions consisting of terms of one letter only, then
the system must be the logical product system P_5. Otherwise,
the [α,β,γ] discussion of §14 reveals that from $\beta_1(A)$, and
an [A:a] α-function whose normal expansion has a term with
more than one letter in its second expression, the function
A+B:ab can be derived. Since every β-function that satisfies
the [A:a] condition is reducible to first order, it follows
the system must now be

A_2. All α and β-functions that satisfy the
[A:a] condition.

We may write $A_2 = A_4 + R_2$. A_2 is therefore of the second
order.

P_5 and A_2 are thus the only closed $[\alpha,\beta]$ systems that possess the function AB:a+b and satisfy the [A:a] condition. Hence, the only closed $[\alpha,\beta]$ systems possessing the function AB:a+b are P_5, A_2, and C_2.

We turn now to closed $[\alpha,\beta]$ systems possessing the function AB+ab:Ab+aB. This function is the β alternating function of A and B with relevant subset (A,B). It therefore generates the alternating system L_2. Since L_2 possesses all α and β alternating functions, any other closed $[\alpha,\beta]$ system possessing the function AB+ab:Ab+aB cannot consist wholly of alternating functions, and hence, by §16, must possess a second order function with a complete expansion having a one term expression. Now the only complete expansions of this type that correspond to α and β-functions of two variables A, B are AB:Ab+aB+ab, AB+Ab+aB:ab, AB+aB+ab:Ab, AB+Ab+ab:aB, which reduce to AB:a+b, A+B:ab, a+B:Ab, A+b:aB respectively. If in AB+ab:Ab+aB the variables A:a and B:b are replaced by A+B:ab and AB+ab:Ab+aB respectively, the function AB:a+b results. The same is true if A:a and B:b are replaced by a+B:Ab and A:a respectively, and by A+b:aB and B:b respectively. We thus see that every closed $[\alpha,\beta]$ system other than L_2 that has the function AB+ab:Ab+aB has the function AB:a+b. Since AB+ab:Ab+aB does not satisfy the [A:a] condition, it follows from the earlier discussion that the system must then be C_2.

The only closed $[\alpha,\beta]$ systems possessing the function AB+ab:Ab+aB are therefore L_2 and C_2.

Returning to the initial result of this section we thus see that every closed $[\alpha,\beta]$ system other than P_5, A_2, L_2, and C_2 satisfies the [:aa] condition. The duals of C_2 and A_2 are the second order $[\alpha, \gamma]$ systems

C_3. All α and γ-functions.

A_3. All α and γ-functions that satisfy the [A:a] condition.

Note also that $A_3 = A_4 + R_3$. As the dual of an [:aa] function is an [AA:] function, we can state immediately that every

closed [α,γ] system other than S_5, A_3, L_3, and C_3 satisfies the [AA:] condition.

§20

[α] SYSTEMS OF SELF-DUAL FUNCTIONS

     Consider first a closed [α] system, consisting wholly
of self-dual functions, which does not satisfy the [A:a] condition. Again we take some function in the system that fails
to satisfy the [A:a] condition, and choose a term in the first
expression and a term in the second expression of the complete
expansion of the function that lead to this failure, and note
the three fold grouping of the arguments of the function with
respect to their capitalization in those terms;

     term in the first expression : capital small small
     term in the second expression : capital small capital.

As in the preliminary discussion of [α] systems, none of these
groups can be empty. By replacing the arguments of the function
by A,B,C according as they are in the first, second or third
group respectively, the term in the first expression becomes Abc,
in the second, AbC. Since the resulting function of A,B,C is
self-dual, the duals of these terms, aBC and aBc, must be in
the second and first expressions respectively of the complete expansion of this function. As it is also an α-function, it can
only be one of the following two functions with complete expansions as indicated.

     f_1(A,B,C); ABC+Abc+aBc+ABc : abc+aBC+AbC+abC.[42]

     f_2(A,B,C); ABC+Abc+aBc+abC : abc+aBC+AbC+ABc.

---

[42]   This function is Kempe's unsymmetrical resultant [13].   In
view of the result following, Kempe's corresponding theory may
be said to be about all self-dual α-functions. Likewise the
function h(A,B,C), shortly to be met, is Kempe's symmetric resultant. The corresponding theory is therefore that of all self-
dual [A:a] α-functions. On the other hand, triadic-rejection
(see Frink [8], p. 487; also earlier pages) is seen by the discussion concluding the next section to generate all self-dual

Suppose that $f_1(A,B,C)$ is thus obtained. Then $f_1(A,B,C)$ is in the system under discussion, and hence $f_1(C,B,A)$ is also in the system. $f_1(A,B,C)$ has for A-component the function BC+bc+Bc:bC, i.e., B+c:bC; $f_1(C,B,A)$ has for A-component the function BC:Bc+bC+bc, i.e., BC:b+c. Now B+c:bC is a β-function that does not satisfy the [A:a] condition. Hence, by the preliminary [α,β] discussion, it, in conjunction with BC:b+c, generates the complete [α,β] system. C_2. It follows from the theorem of §9 that $f_1(A,B,C)$ and $f_1(C,B,A)$ can generate a function f(A,B,C...I) with arbitrary α or β A-component φ(B,C,...I). Clearly f(A,B,C,...I) must be a self-dual α-function. Now every α-function of A,B,C,...I has for A-component some α or β-function of B,C,...I. Furthermore, the a-component of a self-dual function of A,B,C,...I is easily seen to be the dual of its A-component, so that there can be but one self-dual function of A,B,C,...I with given A-component. It follows that the above functions f(A,B,C...I) constitute all self-dual α-functions of A,B,C...I. The given closed [α] system of self-dual functions must therefore be

D_1. All self-dual α-functions.

Suppose, on the other hand, that $f_2(A,B,C)$ was obtained. $f_2(A,B,C)$ is the α alternating function of A,B,C with relevant subset (A,B,C). It therefore generates L_4, which consists of all α alternating functions. If then the given closed system is not L_4, it cannot consist wholly of alternating functions. It will therefore possess, in accordance with §16, a function $f_3(A,B,C)$ whose A-component has a complete expansion with a one term expression. That this is the only possible conclusion in the present case follows from the fact that a self-dual function of the second order would have two terms in each expression of its complete expansion. Now $f_2(A,B,C)$ has for A-component the function BC+bc:bC+Bc. This, with the A-compon-

functions. Interesting, but outside the scope of the present paper, is [25] by Royce. The sort of diagram given by Kempe, [13], p. 172, for symmetric resultant was invaluable to the writer in originally obtaining the constitution of the various generated systems.

ent of $f_3(A,B,C)$, would generate C_2 by the BC+bc:bC+Bc discussion of the preceding section. Hence, as before, the given system would be D_1.

It therefore follows that L_4 and D_1 are the only closed [α] systems of self-dual functions that do not satisfy the [A:a] condition.

We turn now to closed [α] systems of self-dual functions that do satisfy the [A:a] condition. It is interesting to observe in connection with our preliminary [α] systems discussion that by §8 these systems can also be described as the closed [α] systems that satisfy both the [:aa] and [AA:] condition.

Note first that if a self-dual α-function which satisfies the [A:a] condition has a term with but one small letter in the second expression of its complete expansion, then it must be reducible to first order. For if that small letter be e, the normal [A:a] expansion of the function will have e for a term of its second expression. As the function is self-dual, E will be a term of the first expression of that normal expansion which therefore must be E:e. Now O_1 and R_1, the only closed [α] systems of functions reducible to first order, also happen to consist of self-dual functions that satisfy the [A:a] condition. Every other closed [α] system of the type we are considering must therefore possess a function whose complete expansion has every term with but one small letter in its first expression.

This function cannot be of the first or second order, as the only self-dual α-functions of A and of A,B are A:a, AB+Ab:ab+aB, AB+aB:ab+Ab which, in fact, are reducible to first order. If of the third order, with arguments A,B,C, the function can only be AB+BC+AC:ab+bc+ac, since the first expression of its complete expansion must be ABC+ABc+aBC+AbC. Suppose now that the function is of order n, $n > 3$. Its complete expansion will then have at least one term such that it, and hence also its dual, have at least two capital letters and at least two small letters. From such a pair of dual terms choose the term that is in the first expression of the complete expansion; and in the given function replace each argument that appears as a small letter in that term by A, the remaining arguments be B,C...G respectively. The term in question becomes aBC...G. The remaining terms of the resulting complete expansion which have but one small letter are all derived from terms of the same type in the original complete expansion. Hence the resulting function'of A,B,C,...G, which must be of order n' with $n > n' \geq 3$, also has a complete expansion in which every term with but one small letter is in the first expression. By repetition of this process we must finally obtain the third order function of this type.

Hence, every closed [α] system of self-dual functions that satisfies the [A:a] condition, if not O_1 or R_1, possesses the function AB+BC+AC:ab+bc+ac.

We now prove that this function generates the closed system

D_2. All self-dual α-functions that satisfy the [A:a] condition.

It will then follow that O_1, R_1, and D_2 are the only closed [α] systems of self-dual functions that satisfy the [A:a] condition, and hence that O_1, R_1, D_2, L_4, and D_1 are the only closed [α] systems consisting wholly of self-dual functions.

Let the function AB+BC+AC:ab+bc+ac be symbolized h(A,B, C). Since h(A,B,C) satisfies the conditions defining D_2, and those conditions are iterative, every function generated by h(A,B,C) must be in the resulting closed system D_2. Note that h(A,A,B) admits the expansion A:a. Hence h(A,B,C) can generate every function in D_2 that is reducible to first order. Since a function is identically equal to the first expression of any expansion thereof, we can write $h(T,S_1,S_2) = T(S_1+S_2)+S_1S_2$, and obtain by induction, for $n \geq 2$,

$$h_n(T,S_1,S_2,...S_n) = h(T,S_1,h(T,S_2,...h(T,S_{n-1},S_n)...))$$
$$= T(S_1+S_2+...+S_n)+S_1S_2...S_n.\text{[43]}$$

---

[43]   Cf. Frink [8], p. 482.

Let M be any function in D_2 that is not reducible to first
order, N any function on the same arguments that can be generated by h(A,B,C). We shall call the first expression of the
normal expansion of an [A:a] function the normal form of the
function. Then any term of the normal form of M must have at
least two letters; for if that normal form had a one letter term
E, self-duality would make the normal expansion E:e. Let BD..
..G be any term of the normal form of M that is not wholly
contained in N. As that term has ν letters, $\nu \geq 2$, we can
form the function N', also generated by h(A,B,C), as follows.

$$N' = h \ (N,B,D,\ldots G) = N(B+D+\ldots+G)+BD\ldots G.$$

Since D_2 must also satisfy the [AA:] condition, every term
of the normal form of M has a letter in common with the term
BD...G, and hence is contained in B+C+...+G. It follows that
N' contains every term of the normal form of M that N contains, and, in addition, the term BC...G. By repetition of
this process we thus finally obtain a function N_0, generated
by h(A,B,C), on the same arguments as M, and containing M.
Now turn to the complete expansions of M and N_0. Every term
of the first expression of the complete expansion of M is
then in the first expression of the complete expansion of N_0.
Since M and N_0 are self-dual, these first expressions must
each consist of half the total number of terms of the corresponding complete expansions. The first expressions of the complete expansions of M and N_0, are therefore identical, and
hence M and N_0 are identical. h(A,B,C) can therefore also
generate every function in D_2 that is not reducible to first
order. The closed system generated by h(A,B,C) is therefore
D_2.

The new closed systems D_1 and D_2 found in this section
are easily seen to be of the third order, for they have been
generated by functions of order three, and cannot be generated
by first and second order functions since, as we saw above, first
and second order self-dual α-functions are reducible to first
order, and hence can but generate O_1 or R_1.

§21

[α,δ] SYSTEMS

We first observe that a closed [α,δ] system
must consist wholly of self-dual functions.

For the negative of a non-self-dual δ-function is a non-self-dual α-function. By §18 such an α-function yields either the function A+B:ab or the function AB:a+b. As either of these functions in conjunction with $δ_1$(A) generates the [α, β,γ,δ] system C_1, the closed [α,δ] system cannot have either a non-self-dual α-function or a non-self-dual δ-function.

Secondly, the δ-functions possessed by a closed [α,δ] system are the negatives of the α-functions in the system. For the negative of an α-function is a δ-function, and every δ-function is the negative of that α-function which is the negative of the given δ-function. Now the α-functions in a closed [α,δ] system must themselves constitute a closed [α] system. As, by our first observation, this closed [α] system must consist wholly of self-dual functions, it follows that every closed [α,δ] system consists of the functions in a closed [α] system of self-dual functions and their negatives.

We saw, in the preceding section, that the only closed [α] systems of self-dual functions are O_1, R_1, D_2, L_4, and D_1. Now D_1 consists of all self-dual α-functions. Since every self-dual function is either an α or a δ-function, and the condition of self-duality is iterative, it follows that the functions in D_1 with their negatives constitute the closed [α,δ] system

D_3 All self-dual functions.

D_3 is then the maximal [α,δ] system. As with D_1 it is easily verified that D_3 is of order three.

The functions in O_1, R_1, and L_4 with their negatives respectively constitute the known closed [α,δ] systems O_4, R_4, and L_5. On the other hand, the functions in D_2 with

their negatives do not constitute a closed system. For if in
the function AB+BC+AC:ab+bc+ac, which generates D_2, we replace C by its negative, we obtain the function AB+Bc+Ac:
ab+bC+aC which is an α-function, yet, as it does not satisfy
the [A:a] condition, is not in D_2. Hence, O_4, R_4, L_5,
and D_3 are the only closed [α,δ] systems.

§22

[α], [α,β], AND [α,γ] SYSTEMS; FURTHER DISCUSSION

Since we have derived all closed [α] systems of
self-dual functions, we see from our preliminary discussion of [α], [α,β], and [α,γ] systems that the
derivation of all closed systems of these types, and hence
of all types, will be completed when we have determined
all closed systems that satisfy the [:aa] or the
[AA:] condition.

As these conditions are duals of each other, it will be
sufficient to derive explicitly all closed systems that satisfy
the [:aa] condition.

Since AB:a+b does not satisfy the [:aa] condition, it follows from our preliminary [α] systems
discussion that every closed [α] system that satisfies the [:aa] condition, but does not consist wholly
of self-dual functions, possesses the function A+B:ab.

This useful result is made all the more useful by the following observation. The closed [α] systems of self-dual functions that satisfy the [:aa] condition are at the same time
the closed [α] systems of self-dual functions that satisfy
the [A:a] condition. Hence, by §20, the only closed [α]
systems that satisfy the [:aa] condition and do consist wholly
of self-dual functions are O_1, R_1, and D_2.

$γ_1(A)$ and $δ_1(A)$ do not satisfy the [:aa] condition. Hence, every closed system that satisfies the

[:aa] condition is either an [α] or [α,β] system.

We can therefore classify all closed systems that satisfy the [:aa] condition as follows.

1. [:aa] [α] systems not satisfying the [A:a] condition.
2. [:aa] [α] systems satisfying the [A:a] condition.
3. [:aa] [α,β] systems satisfying the [A:a] condition.
4. [:aa] [α,β] systems not satisfying the [A:a] condition.

This fourfold classification is fundamental both in the present and in the following section.

In the present section the following specialization of the [:aa] condition will be required. A function will be said to satisfy the [:a$_\infty$] condition if all the terms of the second expression of an expansion of the function have a small letter in common.[44] As in §8 it is readily verified first, that the [:a$_\infty$] condition is independent of the particular expansion used, second, that the [:a$_\infty$] condition is iterative. We can therefore also refer to a closed system satisfying the [:a$_\infty$] condition. Clearly, the [:a$_\infty$] condition for function or closed system implies the [:aa] condition.

If the common small letter referred to in the [:a$_\infty$] condition is a, it will be convenient to say the function satisfies the [:a$_\infty$] condition with respect to a. This too is independent of the particular expansion used. Note that for such a function, every term of the complete expansion of the function that involves the capital letter A must be in the first expression of that complete expansion.

Hence, a function that satisfies the [:a$_\infty$] condition with respect to a is determined by its a-component.

In applying the mechanism of a-components to closed sys-

---

[44] A typical and important example is the function A + b : aB shown below to generate all functions satisfying the [:a$_\infty$] condition. Reinterpreted as truth-function of propositions, this function is the familiar 'implication' of Principia Mathematica.

tems that satisfy the [:a$_\infty$] condition, as is thus suggested, the following observation will prove to be indispensable.

> If each function in a given set of generators satisfies the [:a$_\infty$] ·condition with respect to a, then every function that can be generated by those generators through iterative processes in which no replacement is effected upon the variable A also satisfies the [:a$_\infty$] condition with respect to a.

We now take up each of the four cases under which a closed system satisfying the [:aa] condition may be classified, and obtain for each case a result which in this section will yield all closed systems satisfying the [:a$_\infty$] condition, and which at the same time will furnish the necessary starting point for the concluding work of the next section.

1. [:aa] [α] <u>systems not satisfying the</u> [A:a] <u>condition</u>.

As in §20, except for the interchange of A and B, we find that every closed system of this type possesses a function f_1(A, B,C) with the terms aBc and aBC in the first and second expressions respectively of its complete expansion. Since abc must be in the second expression of this complete expansion, the a-component of f_1(A,B,C) is a γ-function that fails to satisfy the [A:a] condition. As the closed system under discussion cannot consist wholly of self-dual functions, it must have the function A+B:ab. It therefore also possesses the function f_2(A, B,C) with expansion A+B+C:abc. f_2(A,B,C) then has for a-component the function B+C:bc. By the dual of the explicitly given preliminary [α,β] discussion we see that these a-components generate the complete [α,γ] system C_3. Hence, by the theorem of §9, f_1(A,B,C) and f_2(A,B,C) can generate a function f(A, B,C...I) with arbitrary α or γ a-component φ(B,C,...I).

Since f_1(A,B,C) has the term aBC in the second expression of its complete expansion, and satisfies the [:aa] condition, it follows that every term of this second expression must have the small letter a in common with the term aBC. That is, f_1(A,B,C) satisfies the [:a$_\infty$] condition with respect to a.

The same is evidently true of $f_2(A,B,C)$. Now by §9 $f(A,B,C,$
...I) can be generated by $f_1(A,B,C)$ and $f_2(A,B,C)$ through
iterative processes in which no replacement is effected upon
the variable A. Hence $f(A,B,C...I)$ also satisfies the $[:a_\infty]$
condition with respect to a. Since every α-function of A,B,
C...I has for a-component some α or γ-function of B,C,...I,
while a function that satisfies the $[:a_\infty]$ condition with respect to a is determined by its a-component, it follows that
$f_1(A,B,C)$ and $f_2(A,B,C)$ can generate every α-function of A,
B,C,...I that satisfies the $[:a_\infty]$ condition with respect to
a. By mere interchange of arguments every α-function of A,B,
C,...I that satisfies the $[:a_\infty]$ condition can then be obtained. Hence $f_1(A,B,C)$ and $f_2(A,B,C)$, which are α-func-
tions satisfying the $[:a_\infty]$ condition, generate the closed
system

$\quad\quad$ F_1^∞. All α-functions satisfying the $[:a_\infty]$
condition.

$\quad\quad$ We can therefore conclude that every closed $[\alpha]$
system that satisfies the $[:aa]$ condition but does
not satisfy the $[A:a]$ condition contains F_1^∞.

$\quad\quad$ 2. $[:aa]$ $[\alpha]$ <u>systems satisfying the</u> $[A:a]$ <u>condition</u>.

This is the case that includes the $[:aa]$ $[\alpha]$ systems of self-
dual functions O_1, R_1, and D_2. Every other closed system
of the type under consideration will then not consist wholly of
self-dual functions, and hence will possess the function A+B:
ab. If the first expression of the normal $[A:a]$ expansion of
each function in such a system consists of terms of one letter,
the system must be a logical sum system, and hence can only be
S_1 or S_2. Otherwise some function in the system will have a
normal expansion with a term of at least two letters in its first
expression. By the definition of a normal expansion, every other
term of this first expression has some letter not in the term in
question. Now in this function replace each argument that is
not involved in that term by A, one of the remaining arguments
by B, the rest by C. The term in question becomes BC, while

every other term of the first expression of the resulting expansion will involve the capital letter A. If then we let the function thus obtained replace the variable .B in the function A+B:ab, the resulting function $f_1(A,B,C)$ will have A+BC for the first expression of its normal expansion, and hence A+BC: ab+ac for its normal expansion.

The system under discussion therefore possesses this function $f_1(A,B,C)$. Since it has the function A+B:ab, it will also possess the function $f_2(A,B,C)$ with expansion A+B+C: abc. $f_1(A,B,C)$ has the a-component BC:b+c, $f_2(A,B,C)$ the a-component B+C:bc. These a-components therefore generate the closed system A_4 consisting of all [A:a] α-functions. Now from their normal expansions we see that the a-component of every [A:a] α-function of A,B,C,...I, which satisfies the [:a$_\infty$] condition with respect to a and which is not reducible to $\alpha_1(A)$ is an [A:a] α-function of B,C,...I. $f_1(A,B,A)$, which admits the expansion A:a, can generate every function of A,B,C,...I reducible to $\alpha_1(A)$. It then easily follows, as in the derivation of F_1^∞ , that $f_1(A,B,C)$ and $f_2(A,B,C)$, which are [A:a] α-functions satisfying the [:a$_\infty$] condition with respect to a, generate the closed system

F_2^∞ . All [A:a] α-functions satisfying the [:a$_\infty$] condition.

We can therefore state that, with the exception of O_1, R_1, D_2, S_1, and S_2, every closed [α] system that satisfies the [:aa] condition and also satisfies the [A:a] condition contains F_2^∞.

3. [:aa] [α,β] <u>systems satisfying the</u> [A:a] <u>condition</u>.

The β-functions to be found in such a system must be reducible to first order, and hence constitute either the closed [β] system O_2 or the closed [β] system R_2. In the first case, the α-functions in the system must be of the first order, and hence the system can only be O_5. In every other case the system consists of the functions in R_2 and the functions in some closed [α] system that satisfies both the [:aa] and [A:a] condition. We therefore merely have to see for each $[\alpha]$ system Σ of this type whether $\Sigma+R_2$ is closed, i.e., whether the functions in Σ together with the functions in R_2 constitute a closed system.

If Σ consists wholly of self-dual functions, it can only be O_1, R_1, or D_2. We have $O_1+R_2 = R_5$, $R_1+R_2 = R_6$. On the other hand D_2+R_2 is not closed. For if in AB+BC+AC:ab+bc+ac, which generates D_2, we replace C by $\beta_1(A)$, we obtain A+B: ab which is an α-function not in D_2. If Σ does not consist wholly of self-dual functions, it either is S_1 or S_2, or else contains F_2^∞. We have $S_1+R_2 = S_3$, $S_2+R_2 = S_4$. Furthermore $F_2^\infty+R_2$ is easily seen to be the closed system

F_3^∞. All $[A:a]$ functions satisfying the $[:a_\infty]$ condition.

We therefore have the following result. Every closed $[\alpha,\beta]$ system that satisfies the $[:aa]$ condition and also satisfies the $[A:a]$ condition, if not O_5, R_5, R_6, S_3, or S_4, contains F_3^∞.

4. $[:aa]$ $[\alpha,\beta]$ <u>systems not satisfying the</u> $[A:a]$ <u>condition.</u>

By the methods of §19 and §20 we see that every closed system of this type possesses a function $f_1(A,B)$ with ab in the first, aB in the second expression of its complete expansion. Both AB and Ab must be in the first expression of this complete expansion, since neither of these terms has a small letter in common with the term aB. Hence $f_1(A,B)$ has the complete expansion AB+Ab+ab:aB, which simplifies to A+b:aB.

By replacing B:b in A+b:aB by A+b:aB we obtain the function A+B:ab.[45] Hence the closed system under consideration possesses the function $f_2(A,B,C)$ with expansion A+B+C: abc. Now $f_1(A,B)$ has the a-component b:B, $f_2(A,B,C)$ the a-component B+C:bc. These a-components generate the complete

---

[45]   Cf. Hilbert-Bernays [9], p. 51.

system C_1 . Hence $f_1(A,B)$ and $f_2(A,B,C)$ can generate a function $f(A,B,C,...I)$ with arbitrary a-component $\phi(B,C,...I)$. As $f_1(A,B)$ and $f_2(A,B,C)$ satisfy the [:a$_\infty$] condition with respect to a, it follows, as in the derivation of F_1^∞ , that they, and hence $f_1(A,B)$ alone, generate the closed system

F_4^∞ . All functions satisfying the [:a$_\infty$] condition.

Hence every closed [α,β] system that satisfies the [:aa] condition but does not satisfy the [A:a] condition contains F_4^∞ .

Every closed system satisfying the [:a$_\infty$] condition also satisfies the [:aa] condition, and hence comes under one of the four cases under which all closed [:aa] systems have been classified. From the definition of F_i^∞ , i = 1, 2, 3, 4, we see that every closed [:a$_\infty$] system coming under case i is contained in F_i^∞ . As all of the closed systems specifically mentioned in the above derivations, with the exception of D_2 , satisfy the [:a$_\infty$] condition, it follows as a partial consequence of the above results that the only closed systems satisfying the [:a$_\sim$] condition are

under case 1, F_1^∞ ,
under case 2, O_1 , R_1 , S_1 , S_2 , F_2^∞ ,
under case 3, O_5 , R_5 , R_6 , S_3 , S_4 , F_3^∞ ,
under case 4, F_4^∞ .

The dual of the [:a$_\infty$] condition, which we call the [A$_\infty$:] condition, is the following. All the terms of the first expression of an expansion of the function have a capital letter in common.

The duals of the four new closed systems found in the present section are then the following.

F_5^∞ . All α-functions satisfying the [A$_\infty$:] condition.
F_6^∞ . All [A:a] α-functions satisfying the [A$_\infty$:] con-

dition.

F_7^∞. All $[A{:}a]$ functions satisfying the $[A_\infty{:}]$ condition.

F_8^∞. All functions satisfying the $[A_\infty{:}]$ condition.

It is unnecessary to state the duals of the general results found above. We but add that F_5^∞, O_1, R_1, P_1, P_2, F_6^∞ are the only closed $[\alpha]$ systems satisfying the $[A_\infty{:}]$ condition, while O_6, R_7, R_8, P_3, P_4, F_7^∞, F_8^∞ are the only closed $[\alpha,\gamma]$ systems satisfying the $[A_\infty{:}]$ condition.

That the eight systems F_i^∞, $i = 1, 2, \ldots, 8$ are distinct is very readily verified. We may note that F_4^∞ contains both F_1^∞ and F_3^∞ each of which contains F_2^∞. Dually for the second set of systems. On the other hand, the common part of any F_i^∞, $i = 1,\ldots 4$, and any F_j^∞, $j = 5,\ldots 8$, would be a closed system satisfying both the $[{:}aa]$ and the $[AA{:}]$ condition, and hence could only be O_1, R_1, or D_2. As D_2 does not satisfy either the $[{:}a_\infty]$ or the $[A_\infty{:}]$ condition, while R_1, which itself contains O_1, is contained in each of the above eight systems, this common part is seen to be R_1, i.e., the system of all α-functions reducible to first order.

Finally we consider the orders of these eight systems. As F_4^∞ was generated by a second order function, it, and hence also F_8^∞, is of the <u>second order</u>. On the other hand, the only α-functions of A, and of A,B satisfying the $[{:}aa]$ condition are $A{:}a$, $AB{+}Ab{:}ab{+}aB$, $AB{+}aB{:}ab{+}Ab$, $AB{+}Ab{+}aB{:}ab$, which are all in S_2. Since S_2 is contained in F_2^∞, and yet is distinct from F_2^∞, it follows that F_1^∞, F_2^∞, F_3^∞, and hence also F_5^∞, F_6^∞, F_7^∞ cannot be generated by first and second order functions. Since they have been shown to be generated by third and lower order functions, it follows that they are of the <u>third order</u>.

§23

[α], [α,β], AND [α,γ] SYSTEMS; CONCLUDING DISCUSSION:

THE EIGHT INFINITE FAMILIES OF SYSTEMS

Having thus determined all closed [:aa] systems that satisfy the [:a$_\infty$] condition, and all closed [AA:] systems
that satisfy the [A$_\infty$:] condition, we turn now to closed [:aa]
systems that do not satisfy the [:a$_\infty$] condition, and closed
[AA:] systems that do not satisfy the [A$_\infty$:] condition.
Again our explicit discussion can be restricted to closed [:aa]
systems. We then observe that the general results of the preceding section, which had for·a partial application the determination of all closed [:a$_\infty$] systems, have for the remainder
of their content the following information about a _closed_ [:aa]
system that does not satisfy the [:a$_\infty$] _condition_.

1. If it is an [α] system that does not satisfy
    the [A:a] condition, it contains F_1^∞, the
    system of all [:a$_\infty$] α-functions.
2. If it is an [α] system other than D_2 that sat-
    isfies the [A:a] condition, it contains F_2^∞,
    the system of all [:a$_\infty$] [A:a] α-functions.
3. If it is an [α,β] system that satisfies the
    [A:a] condition, it contains F_3^∞, the system of
    all [:a$_\infty$] [A:a] functions.
4. If it is an [α,β] system that does not satisfy
    the [A:a] condition, it contains F_3^∞, the system
    of all [:a$_\infty$] functions.

This fourfold result, which corresponds to the fourfold
classification of closed [:aa] systems when that classification is restricted to systems not satisfying the [:a$_\infty$] condition, is fundamental in the following development. Note
that the [:a$_\infty$] system contained in a given system as speci-
fied by these results is the largest [:a$_\infty$] system that any
system coming under the same case as the given system could contain, and hence is _the largest_ [:a$_\infty$] _system contained in the_
given system.

Just as the preceding section required that specialization of the [:aa] condition which we termed the [:a$_\infty$] condition, so we now require further specializations of the [:aa] condition.

A function will be said to satisfy the [:a$_m$] condition, m $\geq$ 2, if it admits an expansion with the following property. Any m terms, distinct or not, of the second expression of the expansion have a small letter in common. Note that when m = 2 this condition is the previously symbolized [:aa] condition. Again as in §8, it is readily verified first, that the [:a$_m$] condition is independent of the particular expansion used; second, that the [:a$_m$] condition is iterative.

We can therefore refer to a closed system satisfying the [:a$_m$] condition. From our definition we see that if a function or closed system satisfies the [:a$_m$] condition, it also satisfies the [:a$_{m'}$] condition for every m' with m $\geq$ m' $\geq$ 2.

If a function satisfying the [:a$_m$] condition admits an expansion whose second expression has not more than m terms, those terms will all have a small letter in common, and hence the function also satisfies the [:a$_\infty$] condition. If then a closed system satisfies the [:a$_m$] condition for every finite m, with m $\geq$ 2, every function in the system, and hence the system itself, must satisfy the [:a$_\infty$] condition.

It follows that with every closed system that satisfies the [:aa] condition but not the [:a$_\infty$] condition there is associated a finite integer μ, $\mu \geq$ 2, such that the system satisfies the [:a$_\mu$] condition, but not the [:a$_{\mu+1}$] condition.

We now prove that every closed system that satisfies the [:a$_\mu$] condition, but not the [:a$_{\mu+1}$] condition, possesses the function h(A,B,...E) of order μ+1 with the following complete expansion. The first expression has all complete terms on A,B,...E that

involve more than one capital letter, the second all
complete terms on A,B,...E with one or no capital
letter.

This result is immediate for the exceptional system D_2,
the only closed system of self-dual functions that satisfies the
[:aa], but not the [:a$_\infty$] condition. For μ is then 2,
and the above function is the h(A,B,C) of §20 which generates
D_2. In every other case we can therefore use the fourfold result stated above. The proof for all systems other than D_2
will then take the following form. The system in question must
possess a function f(A,B,...I) that satisfies the [:a$_\mu$] but
not the [:a$_{\mu+1}$] condition. Corresponding to A,B,...I, the
arguments of this function, we define functions M,N,...W, each
of which is a function of the $\mu+1$ variables A,B,...E, such
that f(M,N,...W), considered as a function of A,B,...E, is
is the desired function h(A,B,...E). By proving M,N,...W to
be in the closed system under discussion, we complete the proof
that f(M,N,...W), i.e., h(A,B,...E) is in the system.

f(A,B,...I) will have $\mu+1$ terms $t_1, t_2, \ldots t_{\mu+1}$ in the
second expression of its complete expansion such that any μ
of them have a small letter in common, but all do not. With
these $\mu+1$ terms we set in 1-1 correspondence the $\mu+1$ complete terms $\tau_1, \tau_2, \ldots \tau_{\mu+1}$, on the $\mu+1$ letters A,B,...E,
which involve but one capital letter. When f(A,B,...I) is an
α-function, our definition of the functions M,N,...W that are
to replace A,B,...I respectively then takes the following form.
Every complete term on A,B,...E with more than one capital is
to be in the first expression of the complete expansion of each
function M,N,...W, the term ab...e with no capital is to be
in the second expression of each complete expansion; on the other
hand, a term τ_1 with but one capital is to be in the first or
second expression of the complete expansion of the function M,
N,...W according as the variable A,B,...I to be replaced by
that function is capital or small in the term t_1. The complete
terms on A,B,...E with more than one capital will then be contained in MN...W, and hence will be in the first expression of
the complete expansion of f(M,N,...W) in terms of A,B,...E.
The term ab...e will be contained in mn...w, and hence will

be in the second expression of that complete expansion. Further-more, every term τ_1 will be contained in that complete term on M,N,...W which arises from t_1 when A,B,...I is replaced by M,N,...W respectively. As each t_1 is in the second ex-pression of the complete expansion of f(A,B,...I), each τ_1 will be in the second expression of the complete expansion of f(M,N,...W) in terms of A,B,...E. If then we set f(M,N,...W) = h(A,B,...E), h(A,B,...E) so defined has the desired complete expansion.

To complete our proof when f(A,B,...I) is an α-function, we need only show that M,N,...W are in the system under dis-cussion. Now none of the variables A,B,...I can be small in each of the $\mu+1$ terms $t_1, t_2, \ldots t_{\mu+1}$, as those terms do not have a small letter in common. Hence none of the functions M, N,...W can have all of the terms $\tau_1, \tau_2, \ldots \tau_{\mu+1}$ in the second expression of its complete expansion. Since each variable A,B, ...E is capital in but one of the terms $\tau_1, \tau_2, \ldots \tau_{\mu+1}$, any $\nu < \mu+1$ of those terms do have a small letter in common. It follows that M,N,...W all satisfy the [:a_∞] condition. Clearly each of these functions is an α-function. Since no com-plete term on A,B,...E other than τ_1 and ab...e can have every letter small that is small in τ_1, each of these functions is also seen to satisfy the [A:a] condition. That is, M,N,... W are [:a_∞][A:a]α-functions, and hence belong to each of the systems $F_1^\infty, F_2^\infty, F_3^\infty, F_4^\infty$. As the system under consideration con-tains at least one of these systems, it possesses each of the functions M,N,...W, and hence also the function h(A,B,...E).

This proof requires but a slight modification when f(A,B, ...I) is a β-function. We can no longer place the term ab...e in the second expression of the complete expansion of each func-tion M,N,...W, for, being contained in mn...w, ab...e would then be in the first expression of the complete expansion of f(M,N,...W) in terms of A,B,...E. We therefore choose any term t_0 in the second expression of the complete expansion of f(A,B,...I), and place ab...e in the first or second expres-sion of the complete expansion of M,N,...W according as the variable A,B,...I to be replaced by that function is capital or small in t_0. The term ab...e will then be contained in that complete term on M,N,...W which arises from t_0 when

A,B,...I is replaced by M,N,...W respectively, and hence will
be in the second expression of the complete expansion of f(M,N,
...W) in terms of A,B,...E. By otherwise defining M,N,...W
as when f(A,B,...I) is an α-function, f(M,N,...W) again has
the desired complete expansion in terms of A,B,...E, and hence
defines the function h(A,B,...E). While M,N,...W will no
longer all be [A:a] α-functions, they still satisfy the [:a$_\infty$]
condition, and hence belong to F$_4^\infty$. As f(A,B,...I) is now a
β-function not reducible to first order, the given system must
come under case 4, and hence contains F$_4^\infty$. It therefore again
possesses each of the functions M,N,...W, and hence also the
function h(A,B,...E).

The second expression of the complete expansion of h(A,B,
...E) is

$$Ab...e + aB...e + ... + ab...E + ab...e.$$

As in our previous discussion it easily follows that h(A,
B,...E) is an [A:a] α-function that satisfies the [:a$_\mu$],
but not the [:a$_{\mu+1}$] condition. That it satisfies the [A:a]
condition can also be seen from the fact that the above second
expression can be simplified to

$$bc...e + ac...e + ... + ab...d,$$

i.e., the logical sum of the logical products of the $\mu+1$ small
letters a,b,...e taken μ at a time. This second expression
of the normal expansion of h(A,B,...E) is the most useful in
the following derivations.

Note and recall that every closed system which satisfies
the [:aa], but not the [:a$_\infty$], condition comes under one and
only one of the four cases under which all closed [:aa] systems have been classified, and has associated with it a unique
integer μ, $\mu \geq 2$, such that the system satisfies the [:a$_\mu$],
but not the [:a$_{\mu+1}$], condition. We now prove that, with the
exception of D$_2$, there exists one and only one closed system
for each case i, i = 1, 2, 3, 4, and associated integer μ,
$\mu \geq 2$. This closed system will be designated F$_i^\mu$. That there
is at least one such system can be verified immediately. In fact
define F$_i^\mu$ as follows.

$\underline{\mu \geq 2}$:

F_1^μ. All α-functions satisfying the $[:a_\mu]$ condition.

F_2^μ. All $[A:a]$ α-functions satisfying the $[:a_\mu]$ condition.

F_3^μ. All $[A:a]$ functions satisfying the $[:a_\mu]$ condition.

F_4^μ. All functions satisfying the $[:a_\mu]$ condition.

We then first observe that each system F_1^μ is closed, since it is defined by means of iterative conditions. From its definition we see that F_1^μ possesses the function $h(A,B,\ldots E)$ of order $\mu+1$ which fails to satisfy the $[:a_{\mu+1}]$ condition. Hence F_1^μ, while satisfying the $[:a_\mu]$ condition, does not satisfy the $[:a_{\mu+1}]$ condition, and consequently has μ for its associated integer. Finally, F_1^μ belongs to case 1 since F_1^∞ is evidently the largest $[:a_\infty]$ system that it contains.

We now prove that F_1^μ is the only closed system belonging to case 1, and associated number μ, with the sole exception $i = 2$, $\mu = 2$, for which there is also the closed system D_2. To do so we examine each case in turn.

1. A closed system with associated number μ that comes under case 1 can now be described as a closed $[\alpha]$ system that satisfies the $[:a_\mu]$ condition, contains F_1^∞, and possesses the function $h(A,B,\ldots E)$ of order $\mu+1$. Let $f(A,B,\ldots I)$ be any α-function of $A,B,\ldots I$ that satisfies the $[:a_\mu]$ condition, and let n be the number of terms, apart from $ab\ldots i$, that are in the second expression of the complete expansion of $f(A,B,\ldots I)$. When $n \leq \mu$, these n terms have a small letter in common as a consequence of $f(A,B,\ldots I)$ satisfying the $[:a_\mu]$ condition, and hence $f(A,B,\ldots I)$ satisfies the $[:a_\infty]$ condition. That is, when $n \leq \mu$, $f(A,B,\ldots I)$ is in F_1^∞, and hence in the given system. We now prove inductively that when $n > \mu$, $f(A,B,\ldots I)$ will also be in the system. Assume that all $[:a_\mu]$ α-functions of $A,B,\ldots I$ with $n = m$, $m \geq \mu$ are in the system, and let $f(A,B,\ldots I)$ have $n = m+1$. As $m+1 \geq \mu+1$, we can choose $\mu+1$ distinct terms other than $ab\ldots i$ from the

second expression of the complete expansion of f(A,B,...I).
Now form the μ+1 expressions obtainable from this second expression by arbitrarily omitting one of those μ+1 terms. By
considering those μ+1 expressions to be the second expressions
of complete expansions on A,B,...I, they serve to define μ+1
functions M,N,...Q of A,B,...I. These functions are evidently [:a$_\mu$] α -functions of A,B,...I with m = n. Hence, by
our assumption, they are in the given system, and consequently
h(M,N,...Q) is in the system. Now the second expression of
the complete expansion of h(M,N,...Q) in terms of A,B,...I
consists of those terms which are in all, or all but one, of
the second expressions of the complete expansions of M,N,...Q.
As every term in the second expression of the complete expansion of f(A,B,...I) is at most missing from one of the second
expressions of the complete expansions of M,N,...Q, while the
latter second expressions have no terms that are not in the former, it follows that h(M,N,...Q) and f(A,B,...I) have complete expansions with identical second expressions, and hence
are themselves identical. Since h(M,N,...Q) is in the given
system, we thus see that f(A,B,...I) is in the system. That
is, if all [:a$_\mu$] α-functions of A,B,...I with n = m, m $\geq$ μ,
are in the given system, the same is true for n = m+1. As this
hypothesis was verified for m = μ, our result follows. The
given system therefore possesses all [:a$_\mu$] α-functions, and
being an [:a$_\mu$] [α] system, can only be F_1^μ.

2. The fact that a closed system coming under case 2 satisfies the [A:a] condition suggests using normal, instead of
complete, expansions. If this be done, and exception is made of
D$_2$, the above derivation can be used here with but few modifi-
cations. The induction is now carried through with respect to
the number of terms n in the second expression of the normal
expansion of f(A,B,...I). F_2^∞ takes the place of F_1^∞ as
the starting point of this induction. Finally, in obtaining the
second expression of the normal expansion of h(M,N,...Q) in
terms of A,B,...I we first obtain an expression consisting of
all the terms in the second expression of the normal expansion
of f(A,B,...I), and in addition other terms contained in those
terms. As these other terms can then be omitted, it follows

that $h(M,N,\ldots Q)$ and $f(A,B,\ldots I)$ have normal expansions with identical second expressions. We thus see that the system in question must be F_2^μ.

3. As a closed system coming under case 3 is an $[\alpha,\beta]$ system satisfying the $[A:a]$ condition, its β-functions must be reducible to first order. Since the system contains F_3^∞ which itself contains the $[\alpha]$ system F_2^∞, the α-functions in the system must constitute a closed system, other than D_2, which comes under case 2 and has the same associated integer μ as in the given system. This $[\alpha]$ system being then F_2^μ, the given system consists of the functions in F_2^μ and all β-functions reducible to first order, and hence is F_3^μ.

4. The argument of case 1 carries over completely with F_4^∞ in place of F_1^∞, and n as the number of terms in the second expression of the complete expansion of $f(A,B,\ldots I)$, to show that the system must be F_4^μ.

We have therefore complete proved that except for D_2, the only closed systems satisfying the $[:aa]$, but not the $[:a_\infty]$, condition are F_i^μ, $i = 1, 2, 3, 4, \mu = 2, 3, 4, \ldots$. Since F_i^μ belongs to case i, and has μ for its associated integer, it follows that F_i^μ and F_j^ν are distinct unless $i = j$ and $\mu = \nu$. The systems F_i^μ thus constitute four distinct infinite families of systems, one family for each case i. Since the $[:a_\mu]$ condition is implied by the $[:a_{\mu+1}]$ condition, F_i^μ contains $F_i^{\mu+1}$, so that if the systems in each family be arranged in order of increasing μ, they form an infinite sequence with each system containing the following. As, also, a function which satisfies the $[:a_\mu]$ condition for every μ, $\mu \geqq 2$, satisfies the $[:a_\infty]$ condition, and conversely, we see that the common part of all the systems in the i-th family is F_i^∞.

Finally, we prove <u>that each system F_i^μ is of order</u> $\mu+1$. First, F_i^μ can be generated by the $(\mu+1)$-th order function $h(A,B,\ldots E)$ and a set of generators of the system F_i^∞. As the order of F_i^∞ does not exceed three, the order of F_i^μ is not greater than $\mu+1$. Now consider any set of generators of F_i^μ. This set will have to possess a function $f(A,B,\ldots I)$ that satisfies the [:a$_\mu$] but not the [:a$_{\mu+1}$] condition. Let t_1, t_2, ...,$t_{\mu+1}$ be a corresponding set of $\mu+1$ terms in the second expression of the complete expansion of $f(A,B,...I)$ such that all do not have a small letter in common, but any μ of them do. For each such set of μ terms pick out one letter which is small in each term in that set. No two sets can yield the same letter, or that letter would be small in all $\mu+1$ terms. As there are $\mu+1$ such sets of μ terms, $f(A,B,...I)$ must have at least $\mu+1$ arguments. That is, a set of generators of F_1^μ is at least of order $\mu+1$. The order of F_1^μ therefore is $\mu+1$. This result has the added interest of showing that closed systems exist of every finite order.

The dual of the [:a$_m$] condition may be symbolized [A$_m$:] and is to the effect that any m terms, distinct or not, of the first expression of an expansion of a function have a capital letter in common.

We can then state immediately that the only closed systems other than D_2 that satisfy the [AA:], but not the [A$_\infty$:] condition are the following.

$\underline{\mu \geq 2}$:

F_5^μ. All α-functions satisfying the [A$_\mu$:] condition.

F_6^μ. All [A:a] α-functions satisfying the [A$_\mu$:] condition.

F_7^μ. All [A:a] functions satisfying the [A$_\mu$:] condition.

F_8^μ. All functions satisfying the [A$_\mu$:] condition.

We also have immediately that each of these systems with indicated μ is of <u>order $\mu+1$</u>, and that all the differently symbolized systems thus represented are distinct. They, too, therefore, form four infinite families of systems with the same abstract structure as the first set of families.

That the systems in this second set of four infinite families are distinct from the systems in the first set of four infinite families will follow from the discussion of the common part of two systems that belong one to an infinite family in

the first set, the other to an infinite family in the second
set. To obtain the most complete result we shall consider F_1^{∞}
to be a member of the i-th family with $\mu = \infty$. Then, as in
the preceding section, we see that the common part in question
can only be 0_1, R_1, or D_2, of which the choice narrows down
to R_1 and D_2. Now D_2 is a closed [A:a] [α] system that
satisfies both the [:aa] and [AA:] condition, i.e., both the
[:a$_2$] and [A$_2$:] condition. It is therefore contained in each
F_1^2, and hence is the common part of any F_1^2 and F_j^2 that belong to families in different sets. On the other hand, D_2 cannot be contained in any F_1^{μ} for which $\mu > 2$, since it possesses the function AB+BC+AC:ab+bc+ac which does not satisfy either the [:a$_3$] or [A$_3$:] condition. Hence the common part of
any F_1^{μ} and F_j^{ν} that belong to families in different sets is
but R_1, the system of all α-functions reducible to first order, when either μ or ν is greater than 2.

This concludes our derivation of all closed systems. As a
consequence of the corresponding individual results we can state
that <u>every closed system is of finite order</u>, i.e., can be generated by a finite set of generators. In the preceding sections
only closed systems of the first three orders were disclosed.
The present section, while adding to the third order systems
previously discovered, enables us to state that there are exactly eight closed systems of every finite order greater than three.

§24

SUMMARY

For convenience of reference we here regroup the various
closed systems according to their order, and also index them according to the section in which their definition occurs.

First Order — 9 Systems

$$0_1, \ldots, 0_9.$$

Second Order — 37 Systems

$$R_1,\ldots,R_{13}, \quad S_1,\ldots,S_6, \quad P_1,\ldots,P_6,$$

$$L_1,L_2,L_3, \quad F_4^\infty,F_8^\infty, \quad A_1,A_2,A_3,A_4, \quad C_1,C_2,C_3.$$

Third Order — 20 Systems

$$L_4,L_5, \quad D_1,D_2,D_3, \quad F_1^\infty,F_2^\infty,F_3^\infty,F_5^\infty,F_6^\infty,F_7^\infty,F_1^2,\ldots,F_8^2, \quad C_4.$$

n-th Order, $n > 3,$ — 8 Systems

$$F_1^{n-1},\ldots,F_8^{n-1}$$

. . . .

$$O_1,\ldots,O_9 \quad : \quad §10$$
$$R_1,\ldots,R_{13} \quad : \quad §11$$
$$S_1,\ldots,S_6 \quad : \quad §13$$
$$P_1,\ldots,P_6 \quad : \quad §13$$
$$L_1,\ldots,L_5 \quad : \quad §15$$
$$F_1^\infty,\ldots,F_8^\infty \quad : \quad §22$$
$$\mu = 2,3,\ldots, \quad F_1^\mu,\ldots,F_8^\mu \quad : \quad §23$$

D_1,D_2 : §20,		A_1 : §14		C_1 : §17	
D_3 : §21.		A_2,A_3 : §19		C_2,C_3 : §19	
		A_4 : §18		C_4 : §18	

PART III

CO-ORDINATION AND APPLICATION

§25

THE INCLUSION RELATION[46]

The above determination of all closed systems, i.e., of all iteratively closed membership systems over ν', acquires an added significance when these systems are co-ordinated through the relation of inclusion between systems. Note that system T_1 contains, or as we shall say <u>includes</u> system T_0 if each function in T_0 is also in T_1. If, in addition, T_0 is distinct from T_1, then T_1 <u>properly includes</u> T_0. When the relation of inclusion is restricted to the set of all closed systems in the above sense, it leads to the significant relation of immediate inclusion. A closed system T_1 <u>immediately includes</u> a closed system T_0 if T_1 properly includes T_0, and there is no closed system T such that T_1 properly includes T and T properly includes T_0.

The set of all closed systems, which may also be described as the set of all closed systems included in the complete system C_1, is an infinite set. Consider, on the other hand, a closed system which includes but a finite number of closed systems. Then for such a finite set of closed systems the usually redundant relation of inclusion is completely analysable in terms of the non-redundant relation of immediate inclusion. In this connection we define a <u>finite inclusion chain</u> joining a closed system T_1 to a closed system T_0 as a finite sequence of closed systems $T_1, T_2, \ldots, T_n, T_0$ such that each system in the sequence,

---

[46] Cf. G. Birkhoff [5]. As pointed out in the introduction, even this version of the present section, including footnotes 47 and 48, was written before the recent work on lattices.

except the last, immediately includes the next. It is then easily proved, and in a purely abstract manner, that if T_0 and
T_1 are any two closed systems in a finite set of closed systems
of the above kind, then, if T_1 properly includes T_0, there
is at least one finite inclusion chain joining T_1 to T_0.[47]

The same result holds for two closed systems T_0, T_1 in
our infinite set of all closed systems provided there are but a
finite number of closed systems T such that T_1 properly includes T and T properly includes T_0. If, on the other hand,
there are an infinite number of such closed systems T 'between'
T_0 and T_1, this result need no longer be valid. To state our
positive result for this case we extend both the notion of inclusion chain and immediate inclusion.

We first define an <u>open inclusion chain</u> to be an infinite
sequence of closed systems $T_1, T_2, T_3, \ldots$ such that each system
in the sequence immediately includes the next. An open inclusion chain is then said to <u>immediately include</u> a closed system
T' if each system in the open chain includes T', and there is
no closed system T properly including T' which also is included in each system in the open chain. Finally, a well ordered series, of closed systems, of the form $T_1, T_2, T_3, \ldots T'$,
$T'', \ldots T^{(n)}, T_0$ is called a <u>simple transfinite inclusion chain</u>
joining T_1 to T_0 if each system in the series, except the
last, immediately includes the next, and the resulting open inclusion chain $T_1, T_2, T_3, \ldots$ immediately includes T'.

It can then be proved by a direct consideration of the
eight infinite families of closed systems that if a closed system T_1 properly includes a closed system T_0, and there are
an infinite number of closed systems between T_0 and T_1, then
there exists a simple transfinite inclusion chain joining T_1
to T_0. We thus have the following general result.

---

[47] Since, in this finite case, the number of systems in such a
joining chain must be bounded, we may define the <u>interval</u> from
T_1 to T_0 to be the largest value of n, i.e., the largest
number of 'links' in a chain joining T_1 to T_0. Symbolizing
this interval by $I(T_1, T_0)$, we observe that if T_1 properly
includes T, and T properly includes T_0, then
$$I(T_1, T_0) > I(T_1, T) + I(T, T_0),$$
which is reminiscent of the relation between timelike intervals
in the special relativity theory!

If T_0 and T_1 are any two closed systems such
that T_1 properly includes T_0, then there exists
either a finite or simple transfinite inclusion chain
joining T_1 to T_0.[48]

Since the relation of proper inclusion is transitive, we
immediately have the converse of the above result. It follows
that for closed systems the relation of proper inclusion, and
hence of inclusion, will be completely determined if the following two problems be solved.

(a). To determine for every closed system the closed
systems immediately included therein.

(b). To determine for every open inclusion chain the
closed systems immediately included therein.

In this connection it is of fundamental importance that,
as was pointed out in Part II, all of the differently symbolized
closed systems are in fact distinct.

The solution of problem (a) is at worst tedious. We merely
give the discussion for the closed systems immediately included
in the complete system C_1, as this result is required in the
next section. By §10, every closed system associated with a
given closed first order system is included in the maximal closed
system associated with that first order system. There are eight
such maximal closed systems apart from C_1 itself, namely C_4,
R_2, R_3, D_3, C_2, C_3, R_9, A_1, associated with $0_1, 0_2, \ldots 0_8$, respectively. Of these, R_2, R_3, R_9 are included in A_1, C_4 in
C_2. Hence, every closed system which is not an $[\alpha, \beta, \gamma, \delta]$ system is included in at least one of the four closed systems D_3,
C_2, C_3, A_1. On the other hand, every closed $[\alpha, \beta, \gamma, \delta]$ system
other than C_1 is an alternating system, and hence is included

---

[48]   More generally any series of systems with T_1 and T_0 as
first and last elements may be called an inclusion chain joining
T_1 to T_0 if each system in the series includes all those that
follow it, while no system may be inserted in the series without
destroying this property. It can then be proved for our infinite set of systems that every inclusion chain is either a finite
or simple transfinite inclusion chain.

in the complete alternating system L_1.

We thus have that every closed system properly included in C_1 is included in at least one of the five closed systems D_3, C_2, C_3, A_1, L_1. But it can be directly verified that no one of these five systems is properly included in another. It follows that they constitute all of the closed systems immediately included in C_1.

The solution of problem (b) is easily obtained from our knowledge of the eight families of closed systems. It is readily seen that F_1^μ immediately includes $F_1^{\mu+1}$ for every $\mu \geq 2$, i = 1,2,...8. Hence the eight infinite families give rise to eight open inclusion chains F_1^2, F_1^3, F_1^4, ..., i = 1,2,...8. Since F_1^∞ is the logical product of all the systems in the i-th family, it follows that F_1^∞ is immediately included in the open inclusion chain corresponding to the i-th family, and, in fact, is the only closed system immediately included therein. Now consider any open inclusion chain. As there are but a finite number of closed systems outside of the infinite families, it follows that from a certain system onward all of the systems in the chain belong to the infinite families. Furthermore, it is readily seen that as we pass down the chain, the systems can change from one family to another in only the following ways: from the fourth family to the first or third, from the first or third to the second; and dually for the other four families. Consequently, along the given chain the systems can change from one family to another at most twice, so that from a certain system in the chain onward all of the systems belong to the same family. If this family is the i-th, F_1^∞ is the only closed system immediately included in the given open inclusion chain.

The eight statements 'F_1^2, F_1^3, F_1^4, ... immediately includes F_1^∞,' i = 1,2,...8, thus constitute a complete non-redundant solution of problem (b), since the analysis just given shows that the corresponding result for an arbitrary open inclusion chain can be deduced from these statements.

The analysis of the inclusion relation for closed systems thus achieved is presented pictorially in the accompanying <u>inclusion diagram</u>.[49] Dual closed systems are represented by points symmetrically placed with reference to a central vertical line, self-dual closed systems by points on this line. Due to the inherent difficulties of this requirement of symmetry, only those self-dual closed systems which immediately include, or are immediately included in non-self-dual closed systems are represented in the main diagram, while all self-dual closed systems are separately represented in the supplementary diagram at the left. For each pair of closed systems T_0, T_1 such that T_1 immediately includes T_0 we have joined the corresponding points by a straight line. The systems have been so arranged vertically, that this straight line always slopes downward from T_1 to T_0. In the case of two self-dual closed systems T_0, T_1 the relationship of immediate inclusion is thus represented only in the supplementary diagram. The eight dotted vertical lines serve the double purpose of indicating that the relation of immediate inclusion explicitly represented for systems in the infinite families with $\mu = 2,3,4$ be repeated indefinitely, and that the open inclusion chain F_i^2, F_i^3, F_i^4, ..., $i = 1,2,...8$, thus resulting immediately includes F_i^∞. Now let the supplementary diagram be considered to form part of the main diagram, let all intersections at points other than those representing closed systems be disregarded, and let the diagram be sufficiently extended in the manner indicated by the dotted lines to include all closed systems with μ not exceeding those of the closed systems about to be mentioned, assuming that the latter belong to the infinite families. Then our result about inclusion chains has the following diagrammatic equivalent. A closed system T_1 properly includes a closed system T_0 when and only when there is a broken line, which may have a dotted segment, joining T_1 to T_0 which slopes downward at each point as it runs from T_1 to T_0. The inclusion diagram thus completely determines the inclusion relation for closed systems.

The inclusion diagram can be put to a number of uses. First it can help determine the system generated by a given finite set of generators.[50] Note, either as a consequence of our chain re-

---

[49]   Cf. G. Birkhoff [4], p. 436.

[50]   And thus easily enable one to recover much specific infor-

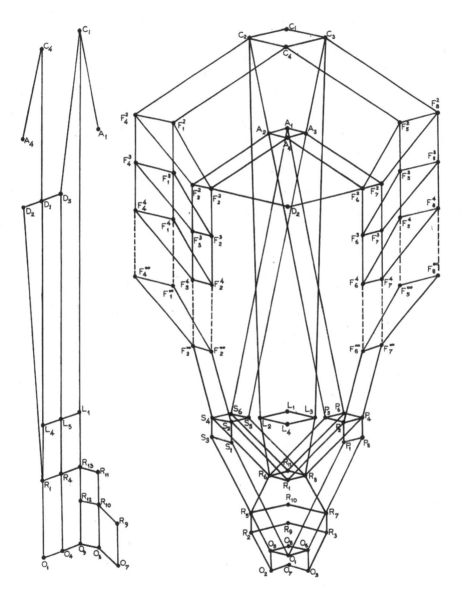

THE INCLUSION DIAGRAM

sults, or by direct proof, that if T_0 is properly included in T_1 then it is included in some closed system immediately included in T_1. Now the formulas given in Part II for the closed systems enable us to determine whether a given function is or is not in a given system. We may therefore first determine whether the given set of generators is contained in some closed system T_1 immediately included in C_1. If not, the generated system must be C_1. Otherwise, we proceed in like fashion with respect to the systems immediately included in T_1, and so on. This process must terminate unless it leads to some system F_1^μ with $\mu > 2$. In that case we pause to see if the given set of generators is contained in F_1^∞. If not, the original process will still terminate. Otherwise the process is re-begun with respect to F_1^∞, and must again terminate.

Another application of the diagram is found in the determination of the logical product and 'iterative sum' of a given set of closed systems. The logical product of a set of closed systems consists of the functions that are common to all of the systems of the set. Hence, if not null, this logical product itself constitutes a closed system. For an iterative process involving only functions which are in each of the given closed systems must yield a function which is also in each of the given systems. By the iterative sum of a given set of closed systems we mean that closed system which is generated by the logical sum of the given systems, i.e., by a set of generators, usually infinite in number, which consists of all of the functions in all of the given systems. Unlike the logical product, the iterative sum always exists for our closed systems.[51]

---

mation which formed an integral part of our original derivation of all closed systems of truth-tables, but was not required by the present derivation; e.g., the closed systems generated by the several second order functions taken one at a time, two at a time, etc.

[51]   Our chief reason for omitting a null closed system was to emphasize the difference between our inclusion diagram, and the inclusion diagram for the subgroups of a group, where an existent subgroup, the identity, is always common to all the subgroups. Were this omission supplied, then what we term below

Assuming their existence, we easily see that the logical product and iterative sum thus defined possess the following inclusion properties.

1. The logical product of a given set of closed systems is a closed system included in each of the given systems and including every closed system included in each of the given systems.

2. The iterative sum of a given set of closed systems is a closed system including each of the given systems and included in every closed system including each of the given systems.

In the case of the logical product the prerequired existence depends on the condition

a. There exists a closed system included in each of the given systems.

In the case of the iterative sum this existence follows from the property

b. There exists a closed system including each of the given systems;

i.e., C_1. Clearly, properties 1 and 2 uniquely determine the closed system having that property. Hence, properties 1 and 2, as qualified by condition a and property b respectively, completely characterize the logical product and iterative sum in

---

an abstract (iterative) inclusion set is identical with a complete lattice ([5], p. 795). However, the writer is not convinced that the formulation in the present section is not to be preferred. Thus, in projective geometry, to preserve the relation $m+n = \mu + \nu$ between the dimensionality of two given spaces and that of their join and meet, it would be necessary to introduce ideal spaces of different negative dimensionalit, for certain different cases of non-intersection, instead of fitting all non-intersections into the one mould of a common null-set as meet.

terms of the inclusion relation. They therefore enable us to
determine the logical product and iterative sum of a given set
of closed systems solely through the use of the inclusion diagram.

In fact, if the set of all closed systems be considered as
an _abstract inclusion set_, i.e., a set of undefined 'closed systems' subject to the present inclusion relation, then 1 and 2
serve to define the logical product and iterative sum of a given
set of closed systems. In that case, the fact that there exists
a closed system having property 1 whenever condition a is satisfied, and that there always exists a closed system with property
2 as a consequence of b being satisfied, constitutes an important property of our inclusion set which serves to distinguish
it from non-iterative inclusion sets. In this connection, however, note that 1 and 2 are not independent. For even with logical product and iterative sum thus abstractly defined, it is
still readily verified that the iterative sum of a given set of
closed systems is the logical product of all of the closed systems which include each of the given systems, while the logical
product of a given set of closed systems is the iterative sum
of all of the closed systems included in each of the given systems. That is, the existence of a closed system with property
2 is a consequence of 1 and b, while the existence of a closed
system with property 1, subject to a, is a consequence of 2.

By the use of the iterative sum our previously described
method of determining the closed system generated by a given set
of generators can be modified as follows. We first determine by
the original method the closed systems separately generated by
each function in the given set of generators. Then the iterative sum of these closed systems will be the closed system generated by the entire set of generators. This modified method is
very useful in the inverse problem of determining all the ways
in which a given closed system can be generated by a set of independent generators. The solution of this problem for the complete system, given in the next section, constitutes our final
application of the inclusion diagram.[52]

---

[52] The solution of this problem is of greater significance
for propositional calculi than for Boolean algebras. For in

§26

SETS OF INDEPENDENT GENERATORS OF THE COMPLETE SYSTEM

We have proved in the preceding section that there are exactly five closed systems immediately included in the complete
system, namely C_2, C_3, D_3, A_1, and L_1. This proof, furthermore, reveals that every closed system other than the complete
system is included in at least one of these five systems. It
follows that the closed system generated by a given set of generators is the complete system when and only when each of the
five systems C_2, C_3, D_3, A_1, and L_1 fails to contain the
given set of generators.

If C_2 is not to contain a given set of generators, at
least one of these generators must be a γ or a δ-function.
Likewise, if C_3 is not to contain the given set of generators,
at least one of these generators must be a β or a δ-function.

We thus immediately arrive at the important result that every set of generators of the complete system possesses either a δ-function, or both a β and
a γ-function.

Either of these possibilities automatically prevents C_2

---

the latter, the primitive membership functions, using the logic
of classes interpretation, may not by themselves generate all
membership functions, though all membership functions may be defined with their help via the general logic in which a Boolean
algebra is immersed. The reader may well feel that we over-
stress the condition that the complete system be generated by
a set of independent generators. Thus, greater symmetry results
if negation, disjunction, and conjunction are used as primitives
instead of merely the first two. Without the condition of independence the problem becomes much easier, and is thus solved
in the beginning of the next section. We should note that the
method of solution suggested above is rather the method used in
our original version of the present paper. In the present development a somewhat different method yields a much neater solution of the general problem of independent generators of the
complete system. On the other hand, the specialization treated
at the end of the next section was more easily arrived at via
the original plan.

and C_3 from containing the given set of generators. Of the
remaining three systems immediately included in the complete
system, D_3 consists of all self-dual functions, A_1 of all
[A:a] functions, L_1 of all alternating functions. Now every
δ-function fails to satisfy the [A:a] condition, while every
β and γ-function is non-self-dual. We can therefore restate
the necessary and sufficient condition that a set of generators
generates the complete system as follows.

      If the set of generators possesses a δ-function,
at least one generator must be non-self-dual and at
least one non-alternating; if the set of generators
possesses both a β and a γ-function, at least one
generator must be a non-[A:a] function and at least
one generator must be non-alternating.

      This condition, in conjunction with the result of the pre-
ceding paragraph, may be considered a solution of the problem
of completely characterizing sets of generators of the complete
system.

      A set of generators of the complete system will clearly
consist of independent generators when and only when no proper
subset of the given set of generators is capable of generating
the complete system.[53] This added restriction on a set of generators of the complete system results in a drastic limitation
upon the number of generators in the set. In fact, if a set of
generators of the complete system possesses a δ-function, by
choosing from this set of generators the δ-function, one non-
self-dual function, and one non-alternating function, we obtain
a subset of the given set of generators which is capable of generating the complete system, and which consists of the δ-function and at most two other generators.

      It follows that if a set of independent generators
of the complete system possesses a δ-function, it can

---

[53]   The definition of independence, of course, being that no
generator in the set can be generated by the remaining generators.

have no more than two other generators. Likewise, if
a set of independent generators of the complete system
possesses both a β and a γ-function, it can have no
more than two other generators.

As a result of this limitation on the number of generators
in a set, it becomes possible to present a set of formulas which
represent all sets of independent generators of the complete system. To achieve conciseness of statement, we introduce the following symbolism.

D : self-dual,	d : non-self-dual,
A : [A:a],	a : non-[A:a],
L : alternating	l : non-alternating,
R : reducible to first order,	r : not reducible to first order.

By D1δ, for example, we shall then mean any self-dual non-
alternating δ-function.

In terms of this symbolism the proof that follows
establishes the result that every set of independent
generators of the complete system is a set of functions
of one of the following forms, and conversely.

$$I : \{d\delta\}$$
$$II : \{D\delta, d\alpha\}$$

III : {D1δ,β}	IV : {D1δ,γ}
V : {Dδ,1β}	VI : {Dδ,1γ}
VII : {Lδ,Lβ,D1α}	VIII : {Lδ,Lγ,D1α}

. . . .

IX : {1β,γ}	X : {β,1γ}

$$XI : \{L\beta, L\gamma, a1\alpha\}$$

XII : {Lrβ,Lγ,1α}	XIII : {Lβ,Lrγ,1α}

$$XIV : \{R\beta, R\gamma, Ar\alpha, Lr\alpha\}.$$

We might also interpret D1δ as the class of all self-dual
non-alternating δ-functions. Then every set of independent generators of the complete system would be a selection from one of

the above sets of classes, and conversely. These sets of classes are not mutually exclusive, i.e., certain sets have selections in common. There is no difficulty in making them mutually
exclusive. But a certain uniqueness which could be shown to be
possessed by the given solution would thereby be lost.

The proof of the above results divides itself into two parts
according as there is, or is not, a δ-function among the generators. In the first of these two cases frequent reference is
made to the following previously demonstrated properties.

1. Every α and δ alternating function is self-dual.
2. Every β and γ-function is non-self-dual.

We recall that in this case there can be no more than two generators besides the assumed δ-function.

If this δ-function is to be the only generator, it must be
both non-self-dual and non-alternating. Hence, by 1, it can be
any non-self-dual δ-function. We thus have I.

When there is just one generator besides the δ-function
in question, this δ-function must be self-dual or it alone
would generate the complete system under I, and the two generators would not be independent. The second generator therefore
has to be non-self-dual and hence, for the same reason, cannot
be assumed to be a δ-function. If this non-self-dual generator
is an α-function, it is also non-alternating by 1, and so immediately gives II. Otherwise, by 2, the non-self-dual generator can be any β or γ-function. But then either the δ-function, or this β or γ-function must be assumed to be non-alternating. Hence III, IV, V, and VI.

Finally, there may be two generators besides the δ-function. If this set of generators is to generate the complete
system, there must be among its members the δ-function, a non-
self-dual function, and a non-alternating function. For independence, these three generators must be distinct, or they would
constitute a proper subset of the given set capable of generating the complete system. Hence the δ-function has to be both
self-dual and alternating, one of the other two generators must
be non-self-dual and alternating, the other non-alternating and
self-dual. By 1, the non-self-dual alternating function can only

be a β or γ-function, while by 2, and the fact that a δ-gen-
erator must in the present instance be alternating, the self-dual
non-alternating function has to be an α-function. These conditions are easily simplified by 1 and 2 to yield VII and VIII.

When a set of independent generators of the complete system
does not possess a δ-function, it must have both a β and a
γ-function, and may have at most two other generators. We also
recall that in this case at least one generator is to be a non-
[A:a] function, and at least one non-alternating. The properties now useful are the following.

1. Every α, β, and γ-function reducible to first
    order is both [A:a] and alternating.
2. Every [A:a] β and γ-function is reducible to
    first order.
3. Every [A:a] alternating α-function is reducible
    to first order.

If there are to be no generators besides the β and γ-
function, at least one of these two functions must be non-alternating. By 1, this non-alternating β or γ-function is not
reducible to first order, and hence, by 2, is also a non-[A:a]
function. Thus IX and X.

When there is but one generator besides the β and γ-
functions, both the β and γ-function must be alternating to
avoid cases IX and X. The remaining generator therefore has to
be non-alternating, and consequently can only be allowed to be
an α-function. Finally, one of the three generators must be
non-[A:a]. If the α-function is non-[A:a], we immediately have
XI. On the other hand, the β or γ-function being non-[A:a]
is equivalent by 1 and 2 to its not being reducible to first order, so that we then have XII and XIII.

There thus remains the case where there are two generators
besides the β and γ-functions. As in the corresponding δ-
function case, the β and γ-functions must be both [A:a] and
alternating, one of the remaining generators non-alternating and
[A:a], the other non-[A:a] and alternating. The last two generators must therefore both be restricted to be α-functions.
With the conditions on the β and γ-generators transformed by

1 and 2, and those on the α-generators by 1 and 3, we easily obtain XIV.

Before passing on to a certain specialization of the above general solution, we pause to note that the concept of the φ-subgroup of a group extends to our closed systems. More generally, consider an iteratively closed system S in the sense of §1, and suppose that S and each of its closed subsystems can be generated by a finite number of generators (as we have seen to be true of our closed systems of membership functions). On the one hand, it readily follows that if there be any function in S which cannot be one of a finite set of independent generators of S, the set of all such functions constitutes a proper closed subsystem of S which may then be called the φ-subsystem of S. An important consequence of our hypothesis on S is that each of its proper closed subsystems is included in a closed subsystem immediately included in S. Calling the latter the maximal subsystems of S (not to be confused with our previous use of this phrase), it follows as in finite group theory that the φ-subsystem of S is the cross-cut, i.e., logical product, of the maximal subsystems of S. In particular, the φ-subsystem of the complete system of membership functions C_1 is therefore the cross-cut of C_2, C_3, A_1, D_3, and L_1, and hence, either by the inclusion diagram, or more easily by property 3 above, is seen to be R_1, the system of functions reducible to first order α-functions.[54]

The requirement that a set of generators of the complete system consist of independent generators may be considered to be but a partial statement of the more general requirement that a set of generators of the complete system be not burdened with unnecessary complications. Various standards of irreducible simplicity may be adopted, each leading to a specialization of our general solution. We shall consider but one such specialization.

Only those sets of independent generators of the

---

[54] This paragraph is a recent addition. While it suffices for our present purpose, its extensions and ramifications are manifold. But such a wider treatment would have to be related to the recent developments of lattice theory.

complete system are to be admitted in which one cannot
replace a generator by a function of lower order that
can be generated by this generator, and thus obtain a
new set of independent generators of the complete system.[55]

An important consequence of this restriction, we shall see,
is that it replaces the infinite number of sets of merely independent generators, infinite even when abstraction is made of
the particular variables on which they are written, by a finite
number of sets. Hence, while the solution of the unrestricted
problem can only be given in terms of conditions imposed upon
the generators, the solution of the restricted problem gives us
the various individual sets themselves.

We have seen that every set of independent generators of
the complete system is a selection from at least one of the
sets of classes of functions I — XIV, and conversely. Our
specialization clearly rejects every selection which chooses
from a class a function that can generate a function of lower
order in that class. We must therefore first restrict each
class to those functions that cannot generate a function of
lower order in that class. It will then be easy to see which
of the selections from the resulting sets of <u>restricted classes</u>
also fail to conform to our specialization.

This program can be carried out with far greater ease if
we break up each of the classes $a l\alpha$, $l\alpha$, and $Ar\alpha$ into two
classes as follows:

$$a l\alpha = da\alpha + Dal\alpha, \quad l\alpha = d\alpha + Dl\alpha, \quad Ar\alpha = dA\alpha + DAr\alpha,$$

correspondingly replace each of the sets of classes XI, XII,
XIII, and XIV by two sets of classes, and apply the above plan
to these <u>modified classes and sets of classes</u>. As a result of
this modification, each class now consists entirely either of
self-dual, or of non-self-dual functions.

---

[55] The new set, if a set of generators of the complete system,
will be a set of independent generators.

This simplification enables us to state the uniform result that the restricted classes now consist
of the functions of lowest order in each class.

To prove this result it is clearly sufficient to show for
each class that every function in the class can generate one of
the functions of lowest order in that class. This lowest order
is easily seen to be the first order for the classes Dδ, Lδ, β,
Lβ, Rβ, γ, Lγ, Rγ, the second order for the classes dδ, dα,
1β, 1γ, Lrβ, Lrγ, dAα, and the third order for the remaining
classes Dlδ, Dlα, daα, Dalα, DArα, Lrα. Our result then follows immediately for the Dδ, Lδ, β, Lβ, Rβ, γ, Lγ, and Rγ
classes since they consist of functions whose associated first
order functions are in the same class. In the case of the dα
and dAα classes, §18 shows that every function therein generates at least one of the second order functions A+B:ab, AB:
a+b, both of which are in the classes concerned. A similar argument disposes of the dδ class. By §16, every function in
the Lrβ class generates a second order Lrβ function. Dually
for Lrγ. Each Dlδ, Dlα, Dalα, DArα, and Lrα function generates a closed system which can be generated by a single third
order function. This third order function is therefore in the
same class as the given function, and is generated by the given
function. For the remaining classes we have recourse to the inclusion diagram. This shows that each 1β function generates
a closed system containing F_4^∞, which in turn can be generated
by the 1β function a+B:Ab. Dually for 1γ functions. Likewise each daα function generates a closed system containing
F_1^∞ or F_5^∞, each of which can be generated by a single third
order daα function.

Each restricted class thus consists of functions of a single
order not exceeding three. Before these functions are obtained
explicitly, it is desirable to reconsider the final form we wish
the solution of our specialized problem to take. We certainly
do not wish to distinguish between sets of generators which differ solely in the particular variables used in their expression.
More drastically, we may not wish to distinguish between two
generators which can be obtained from each other by a mere 1-1
replacement of their arguments. If we consider such generators

to correspond to the same <u>abstract function</u>, our problem then takes the form of finding the different <u>abstract sets of generators</u> that correspond to sets of generators agreeing with our specialization. We shall adopt this point of view since it leads to the completest analysis, while if it be further desired to find all sets of generators corresponding to a given abstract set of generators, no new difficulty is encountered. In connection with the restricted classes, this requires us to know the different abstract functions that correspond to the functions therein. In fact, it will be convenient to consider the restricted classes as composed of these abstract functions, and in like manner, to reconceive our entire program in terms of abstract functions. Our main remaining problem therefore is to effect an analysis of abstract functions of the first three orders sufficient to describe the restricted classes from this new point of view.

Each of the four first order functions on a given argument gives rise to a distinct abstract function. Of these four possible abstract first order functions but three are seen to be members of the restricted classes. In terms of our previous symbolism they are β_1, γ_1, and δ_1.

We will readily verify that the second order abstract functions which are members of the restricted classes comprise all second order abstract functions not reducible to first order. In counting these abstract functions, we first observe that of the sixteen second order functions on two arguments A, B, the eight functions which have Ab and aB in opposite expressions of their complete expansions give rise to but four different abstract functions. There are thus but twelve abstract second order functions. Of these, four are reducible to the four first order functions respectively. The number of abstract second order functions not reducible to first order is therefore eight. We symbolize and represent them as follows.

α_2', A+B:ab; β_2', a+B:Ab; β_2'', AB+ab:Ab+aB; δ_2', ab:A+B

α_2'', AB:a+b; γ_2', aB:A+b; γ_2'', aB+Ab:ab+AB; δ_2'', a+b:AB

The third order abstract functions that are members of the restricted classes are perhaps most easily found by an enumeration of all abstract third order α-functions. Consider an arbitrary α-function of A, B, and C. Let the sign + or - be associated with a complete term on A, B, C according as that term is in the first or second expression of the complete expansion of the function. The terms ABC and abc then have the sign + and - respectively in all third order α-functions, and remain unchanged when A, B, and C are permuted. We therefore focus our attention on the remaining terms which we group in three pairs of dual terms as follows, (aBC, Abc), (AbC, aBc), (ABc, abC). A pair of signs can be assigned to each of these pairs of dual terms in four different ways, thus giving rise to the $4 \cdot 4 \cdot 4 = 64$ third order α-functions of A, B, and C. We observe, however, that permuting the variables A, B, C has the effect of correspondingly permuting the three pairs of dual terms. Hence, in counting abstract functions we can disregard the order in which the three pairs of signs are assigned to the three pairs of dual terms. There are thus 4 abstract functions corresponding to the three pairs of signs being all different, $4 \cdot 3 = 12$ abstract functions with two pairs of signs the same, the third different, 4 abstract functions with all three pairs of signs the same. The number of abstract third order α-functions is thus but 20. They may be represented as follows.[56]

	1	2	3	4	5	6	7	8	9	10	11	12	13	14	15	16	17	18	19	20
ABC	+	+	+	+	+	+	+	+	+	+	+	+	+	+	+	+	+	+	+	+
aBC	+	+	+	+	+	+	+	+	+	+	-	-	-	-	-	-	+	+	-	-
AbC	-	-	+	+	+	+	+	+	+	+	-	-	-	-	-	-	+	+	-	-
ABc	-	-	-	-	+	-	-	+	-	-	+	+	-	+	+	-	+	+	-	-
abC	+	+	+	-	+	-	+	-	-	+	-	+	+	-	+	-	-	+	-	+
aBc	-	-	+	+	-	-	-	+	+	+	-	-	-	+	+	+	-	+	-	+
Abc	+	+	-	-	-	-	-	+	+	+	-	-	-	+	+	+	-	+	-	+
abc	-	-	-	-	-	-	-	-	-	-	-	-	-	-	-	-	-	-	-	-

[56] The key to the resulting tabulation is the above ordering of the pairs of dual terms coupled with the following ordering of the pairs of possible signs: +-, ++, --, -+, an ordering in keeping with our ordering α_1, β_1, γ_1, δ_1 of the four first order functions. The correctness of the tabulation can then be checked in a few minutes.

The only abstract third order functions that are members of the restricted classes are readily seen to be self-dual α-functions not reducible to first order, non-self-dual non-[A:a] α-functions, and self-dual non-alternating δ-functions. Of the 20 abstract third order α-functions, 7, 14, 17, and 20 only are self-dual, with 7 reducible to first order.

Hence, 14, 17, and 20, which we resymbolize $\alpha_3^!$, $\alpha_3^"$, and $\alpha_3^{'''}$ respectively, are the only abstract third order self-dual α-functions not reducible to first order. The above tabulation also reveals that there are exactly ten abstract third order non-self-dual non-[A:a] α-functions, to wit, 1, 2, 3, 4, 9, 10, 12, 13, 15, and 16. We resymbolize them $\alpha_3^{(i)}$, $i = 4,5,...13$. Finally, the abstract third order self-dual non-alternating δ-functions are clearly the negatives of the corresponding α-functions. They are therefore the negatives of $\alpha_3^!$ and $\alpha_3^"$, and may be symbolized $\delta_3^!$ and $\delta_3^"$ respectively.

We can now rapidly complete the solution of our specialized problem. With the help of the preceding analysis of abstract functions, and our proved result that the modified restricted classes consist of the functions of lowest order in each class, we readily find these modified restricted classes to have the following membership.

$d\delta$: $\delta_2^!,\delta_2^"$	β : β_1	γ : γ_1	$d\alpha$: $\alpha_2^!,\alpha_2^"$
$Dl\delta$: $\delta_3^!,\delta_3^"$	$l\beta$: $\beta_2^!$	$l\gamma$: $\gamma_2^!$	$Dl\alpha$: $\alpha_3^!,\alpha_3^"$
$D\delta$: δ_1	$L\beta$: β_1	$L\gamma$: γ_1	$da\alpha$: $\alpha_3^{(i)}$, $i=4,5,..13$
$L\delta$: δ_1	$Lr\beta$: $\beta_2^"$	$Lr\gamma$: $\gamma_2^"$	$Dal\alpha$: $\alpha_3^!$
	$R\beta$: β_1	$R\gamma$: γ_1	$dA\alpha$: $\alpha_2^!,\alpha_2^"$
			$DAr\alpha$: $\alpha_3^"$
			$Lr\alpha$: $\alpha_3^{'''}$

The abstract sets of generators agreeing with our specialization will then be among the selections from the sets of restricted classes corresponding to the modified sets of classes
I — XIV. Of these selections, the only ones still not con-
forming with our specialization are easily seen to be $\{\beta_2'', \gamma_1,$
$\alpha_3'\}$ and $\{\beta_1, \gamma_2'', \alpha_3'\}$, where β_2'' and γ_2'' can be replaced by
β_1 and γ_1 respectively.

We thus find that there are exactly thirty-six
abstract sets of independent generators of the complete system that conform with our specialization.

Regrouped according to the original sets of classes I —
XIV, they are as follows.

$$I : \{\delta_2'\}, \{\delta_2''\}$$

$$II : \{\delta_1, \alpha_2'\}, \{\delta_1, \alpha_2''\}$$

$$III : \{\delta_3', \beta_1\}, \{\delta_3'', \beta_1\} \qquad\qquad IV : \{\delta_3', \gamma_1\}, \{\delta_3'', \gamma_1\}$$

$$V : \{\delta_1, \beta_2'\} \qquad\qquad VI : \{\delta_1, \gamma_2'\}$$

$$VII : \{\delta_1, \beta_1, \alpha_3'\}, \{\delta_1, \beta_1, \alpha_3''\} \qquad VIII : \{\delta_1, \gamma_1, \alpha_3'\}, \{\delta_1, \gamma_1, \alpha_3''\}$$

. . . .

$$IX : \{\beta_2', \gamma_1\} \qquad\qquad X : \{\beta_1, \gamma_2'\}$$

$$XI : \{\beta_1, \gamma_1, \alpha_3'\}, \{\beta_1, \gamma_1, \alpha_3^{(1)}\}, \quad 1 = 4, 5, \ldots 13.$$

$$XII : \{\beta_2'', \gamma_1, \alpha_2'\}, \{\beta_2'', \gamma_1, \alpha_2''\}, \qquad XIII : \{\beta_1, \gamma_2'', \alpha_2'\}, \{\beta_1, \gamma_2'', \alpha_2''\},$$

$$\{\beta_2'', \gamma_1, \alpha_3''\} \qquad\qquad \{\beta_1, \gamma_2'', \alpha_3''\}$$

$$XIV : \{\beta_1, \gamma_1, \alpha_2', \alpha_3''\}, \{\beta_1, \gamma_1, \alpha_2'', \alpha_3''\}, \{\beta_1, \gamma_1, \alpha_3', \alpha_3''\}.$$

Our notation immediately tells whether a generator is an
α, β, γ, or δ-function, and, via the subscript of its symbol,
what its order is. We thus see that of these thirty-six sets of

generators twelve are of the second order, twenty-four of the third order. Wernick [36] has observed for functions of two variables that there may be as many as three independent generators of the complete system, but not more than three. We may add that if functions of the first three orders are considered, there may be as many as four independent generators of the complete system, but that there can be no more than four no matter what the orders of the generators are.

Some of the occurrences of these sets of independent generators in the literature may be noted. Interpreted as truth-functions, we may note that $\{\delta_2'\}$ primarily, its dual $\{\delta_2''\}$ secondarily, termed rejection and incompatibility respectively, were introduced by Sheffer [27], the latter being the one used by Nicod [21]. $\{\delta_1, \alpha_2'\}$ is the negation and disjunction of <u>Principia Mathematica</u> [37], the dual set $\{\delta_1, \alpha_2''\}$, negation and conjunction, being employed by Lewis [15] for "material implication". Still earlier, the <u>Principles of Mathematics</u> [26], presumably following Frege, suggested negation and implication, $\{\delta_1, \beta_2'\}$, as primitives. These have since reappeared in the $C - N$ calculus.[57] On the other hand, the recent $C - O$ calculus of Wajsberg [34] may be said to use the set $\{\beta_2', \gamma_1\}$.

As pointed out elsewhere,[58] the applicability of these results to Boolean algebras is rather obscured. However, where an element interpretable as the universal class is postulated, we may perhaps consider it an added primitive; and by identifying that primitive with the constant function β_1, the following additional occurrences may be noted. $\{\beta_1, \gamma_2'\}$ in postulates for 'exception' ([1], [31]), $\{\beta_1, \gamma_2'', \alpha_2''\}$ for "Boolean Rings with Unit", [28],[59] and perhaps $\{\delta_3'', \beta_1\}$ for 'ternary rejec-

---

[57]   e.g., see [34]. Of course, C is β_2', N is δ_1.

[58]   See footnote 52.

[59]   Here, and in other cases, references to numerous papers of B. A. Bernstein other than those appearing in our bibliography could be given.

tion', ([8], [38]). While, as has been observed by B. A. Bernstein [1], one and the same set of postulates for a Boolean algebra will serve both for a given interpretation and its dual,
in the case of propositional calculi the non-self-dual character of assertability requires entirely different postulational
developments for dual sets of primitive truth-functions.

BIBLIOGRAPHY

1. B. A. Bernstein. A Complete Set of Postulates for the
     Logic of Classes Expressed in Terms of the Opera-
     tion "Exception", and a Proof of the Independence
     of a Set of Postulates due to Del Re.
     University of California Publications in Mathema-
     tics, Vol. 1, (1912 - '24), No. 4 (15th May, 1914),
     pp. 87 - 96.

2. B. A. Bernstein. Complete Sets of Representations of Two-
     element Algebras.
     Bull. Amer. Math. Soc., Vol. 30, (1924), pp. 24 - 30.

3. B. A. Bernstein. The Dual of a Logical Expression.
     Ibid., Vol. 33, (1927), pp. 309 - 311.

4. Garrett Birkhoff. On the Structure of Abstract Algebras.
     Proceedings of the Cambridge Philosophical Society,
     Vol. 31, (1935), pp. 433 - 454.

5. Garrett Birkhoff. Lattices and Their Applications.
     Bull. Amer. Math. Soc., Vol. 44, (1938), pp. 793 -
     800.

6. Archie Blake. Canonical Expressions in Boolean Algebra.
     Review of, Journal of Symbolic Logic, Vol. 3, (1938),
     p. 93.

7. Alonzo Church. An Unsolvable Problem of Elementary
     Number Theory.
     Amer. Jour. Math., Vol. 58, (1936), pp. 345 - 363.

8. Orrin Frink. The Operations of Boolean Algebras.
     Annals of Mathematics, 2 s. Vol. 27, (1925 -'26),
     pp. 477 - 490.

9. D. Hilbert and P. Bernays. Grundlagen der Mathematik.
        Vol. 1, Berlin, 1934.

10. D. Hilbert and P. Bernays. Grundlagen der Mathematik.
        Vol. 2, Berlin, 1939: supplement 3.

11. Tomoharu Hirano. Die Kontradiktorische Logik.
        Review of, Journal of Symbolic Logic, Vol. 3, (1938),
        p. 90.

12. W. S. Jevons. Pure Logic.
        London, 1864.

13. A. B. Kempe. On the Relation Between the Logical Theory
        of Classes and the Geometrical Theory of Points.
        Proc. London Math. Soc., Vol. 21, (1889 - '90),
        pp. 147 - 182.

14. Stanislaw Łésniewski. Grundzüge Eines Neuen Systems der
        Grundlagen der Mathematik.
        Fund. Math., Vol. 14, (1929), pp. 1 - 81.

15. C. I. Lewis and C. H. Langford. Symbolic Logic.
        New York, 1932.

16. Eugen Gh. Mihailescu. Recherches sur un Sous-système du
        Calcul des Propositions.
        Review of, Journal of Symbolic Logic, Vol. 2, (1937),
        p. 51.

17. Eugen Gh. Mihailescu. Recherches sur la Négation et
        l'equivalénce dans le Calcul des Propositions.
        Review of, Ibid., Vol. 2, (1937), p. 173.

18. Eugen Gh. Mihailescu. Recherches sur l'equivalénce et la
        Réciprocite dans le Calcul des Propositions.
        Review of, Ibid., Vol. 3, (1938), p. 55.

19. Eugen Gh. Mihailescu. Recherches sur les Formes Normales
        par Rapport à l'equivalénce et la Disjonction, dans
        le Calcul des Propositions.
        Review of, Ibid., Vol. 4, (1939), pp. 91 - 92.

20. Ernest Nagel. Review by.
        Ibid., Vol. 4, (1939), pp. 35 - 36.

21. J. G. P. Nicod. A Reduction in the Number of Primitive
     Propositions of Logic.
     Proceedings of the Cambridge Philosophical Society,
     Vol. 19, (Jan. 1917), pp. 32 - 41.

22. E. L. Post. Determination of all Closed Systems of Truth
     Tables.
     Abstract, Bull. Amer. Math. Soc., Vol. 26, (1920),
     p. 437.

23. E. L. Post. Introduction to a General Theory of Elemen-
     tary Propositions.
     Amer. Jour. Math., Vol. 43, (1921), pp. 163 - 185.

24. W. V. Quine. Review by.
     Journal of Symbolic Logic, Vol. 2, (1937), p. 175.

25. Josiah Royce. An Extension of the Algebra of Logic.
     Jour. Philos. etc., Vol. 10, (1913), pp. 617 - 633.

26. Bertrand Russell. The Principles of Mathematics.
     2nd edition, London, 1937, New York 1938.

27. H. M. Sheffer. A Set of Five Independent Postulates for
     Boolean Algebras, with Application to Logical Con-
     stants.
     Trans. Amer. Math. Soc., Vol. 14, (1913), pp. 481 -
     488.

28. M. H. Stone. The Representation of Boolean Algebras.
     Bull. Amer. Math. Soc., Vol. 44, (1938), pp. 807 -
     816.

29. A. Suschkewitsch. Untersuchungen uber Verallgemeinerte
     Substitutionen.
     Atti-Congresso Bologna, Vol. 2, (1930), pp. 147 -
     157.

30. Alfred Tarski. Sur le Terme Primitif de la Logistique.
     Fund. Math., Vol. 4, (1923), pp. 196 - 200.

31. J. S. Taylor. A Set of Five Postulates for Boolean Alge-
     bras in Terms of the Operation "Exception".
     University of California Publications in Mathematics,
     Vol. 1, (1912 - 24), No. 12 (12th April 1920), pp.
     241 - 248.

32. Oswald Veblen. The Cambridge Colloquium Lectures.
     1916. Part II. (Analysis Situs).

33. Mordechaj Wajsberg. Beiträge zum Metaaussagenkalkul I.
     Monatsh. Math. Phys., Vol. 42, (1935), pp. 221 - 242.

34. Mordechaj Wajsberg. Métalogische Beiträge.
     Review of. Journal of Symbolic Logic, Vol. 2,
     (1937), pp. 93 - 94.

35. Donald L. Webb. The Algebra of n-valued Logic.
     Comptes rendus des séances de la Société des Sci-
     ences et des Lettres de Varsovie, Class III, Vol.
     29, (1936 - '37), pp. 153 - 168.

36. William Wernick. Complete Sets of Logical Functions.
     Abstract, Bull. Amer. Math. Soc., Vol. 46, (1940),
     p. 235.

37. A. N. Whitehead and Bertrand Russell. Principia Mathema-
     tica.
     2nd Edn., Vol. I, Cambridge, England, 1925. Part 1,
     Section A.

38. Albert Whiteman. Postulates for Boolean Algebra in Terms
     of Ternary Rejection.
     Bull. Amer. Math. Soc., Vol. 43, (1937), pp. 293 -
     298.

39. Norbert Wiener. Certain Formal Invariances in Boolean
     Algebras.
     Trans. Amer. Math. Soc., Vol. 18, (1917), pp. 65 - 72.

40. Norbert Wiener. Bilinear Operations Generating all Oper-
     ations Rational in a Domain Ω .
     Annals of Mathematics, 2s., Vol. 21, (1920), pp.
     157 - 165.

41. Norbert Wiener. Certain Iterative Characteristics of Bi-
     linear Operations.
     Bull. Amer. Math. Soc., Vol. 27, (1920), pp. 6 - 9.

42. Eustachy Żyliński. Some Remarks Concerning the Theory of
     Deduction.
     Fund. Math., Vol. 7, (1925), pp. 203 - 209.